Tap

Tap

Unlocking the Mobile Economy

Anindya Ghose

The MIT Press
Cambridge, Massachusetts
London, England

Set in ITC Stone Serif Std by Toppan Best-set Premedia Limited. Printed and bound in the United States of America.

Library of Congress Cataloging-in-Publication Data

Names: Ghose, Anindya, author.
Title: Tap : unlocking the mobile economy / Anindya Ghose.
Description: Cambridge, MA : MIT Press, [2017] | Includes bibliographical references and index.
Identifiers: LCCN 2016043788 | ISBN 9780262036276 (hardcover : alk. paper)
Subjects: LCSH: Mobile commerce. | Cell phone advertising. | Consumer behavior. | World Wide Web--Security measures.
Classification: LCC HF5548.34 .G46 2017 | DDC 381/.142--dc23 LC record available at https://lccn.loc.gov/2016043788

ISBN: 978-0-262-03627-6

10 9 8 7 6 5 4 3 2

Contents

Acknowledgments vii

Introduction 1

I Human Behavior and the Mobile Phone Journey So Far

1 Mobile Phones: A Truly Transformative Technology 17
2 What the Smartphone Has Changed 25
3 Striking a Balance 33

II The Forces Shaping the Mobile Economy

4 Context: What's Going On? 47
5 Location: Why Geography Matters 59
6 Time: It's On Your Side 79
7 Saliency: Can You See Me Now? 95
8 Crowdedness: Why Scarcity of Space Matters 107
9 Trajectory: Watch Where You're Walking 119
10 Social Dynamics: You Are Who You're With 135
11 Weather: Creating the Perfect Storm 147
12 Tech Mix: Solving Wanamaker's Riddle 159

III Next-Generation Technology Forces

13 The Growing Intimacy between Us and Our Devices 181
14 The Next-Generation Technologies 183

Epilogue 195

Notes 205
Index 227

Acknowledgments

I am grateful to my co-authors on the core research that anchors the book. I have learned an extraordinary amount from my interactions with them: Panos Adamopoulos, Michelle Andrews, Jason Chan, Avi Goldfarb, Xitong Guo, Sangpil Han, Hyeokkoo Eric Kwon, Dongwon Lee, Beibei Li, Siyuan Liu, Xueming Luo, Dominik Molitor, Wonseok Oh, Sung-Hyuk Park, Philip Reichart, Martin Spann, Paramvir Singh, Vilma Todri, Kaiquan Xu, Sha Yang, and Zheng Fang. It has been a ton of fun to work with them.

I would like to thank my NYU Stern colleagues as well as current and past doctoral students with whom I have benefited from various formal and informal scholarly interactions and academic collaborations.

I am particularly indebted to Batia Wiesenfeld for being an incredible mentor for the past twelve years at NYU and for always being there for me. I especially thank Mike Pinedo, Alex Tuzhilin, Peter Henry, Elizabeth Morrison, Russ Winer, and Eitan Zemel for their support in recent years. I also thank Arun Sundararajan for sharing important insights over the years that have helped shape my thinking on various fronts.

It has also been wonderful to get to know so many outstanding NYU Stern undergraduate and graduate students from the MBA, MSBA, EMBA, and TRIUM programs.

I am grateful to my PhD advisors, Ramayya Krishnan and Tridas Mukhopadhyay, and to other members of my dissertation committee—Uday Rajan, Michael D. Smith, and Rahul Telang—for guiding me during my four wonderful years at Carnegie Mellon. I especially thank Uday for being the most incredible coach and mentor that any doctoral student can hope for. Most of what I have learned about research was learned from him.

I would also like to thank several others from whom I have benefited through numerous scholarly discussions and collaborations: Sinan Aral,

Susan Athey, Nikolay Archak, Ravi Bapna, Sofia Bapna, Erik Brynjolfsson, Ram Chellappa, V. Choudhary, Chris Forman, Esther Gal-Or, Bin Gu, Ke-Wei Huang, Yan Huang, Amit Mehra, Vallabh Sambamurthy, Ramesh Shankarnarayanan, Yong Tan, Hal Varian, Sunil Wattal, Yuliang (Oliver) Yao, Sam Jeon, Yuqian Xu, Xuan Ye, Qiang Ye, Byungjoon Yoo, and Rong Zheng. During my very memorable stint at the Wharton School in 2011 and 2012, I benefited from my interactions with David Bell, Eric Bradlow, Gerard Cachon, Peter Fader, Lorin Hitt, Kartik Hosanagar, and Raghu Iyengar.

In the industry and corporate world, I have benefited immensely from my interactions with Nic Baddour, Akshay Chaturvedi, John Blankenbaker, Alejandro Cremades, Andrada Comanac, Rahul Guha, Jack Hanlon, Shankar Iyer, Nadia Neytcheva, George Pappachen, Ross Rizley, Craig Stacey, Wally Wang, and Andrew Wong. Of these I am especially indebted to Shankar Iyer for opening for me one of the most fascinating doors in my career so far: the world of data-driven litigation consulting. Working on these litigation cases has significantly enhanced the breadth and depth of my understanding of the digital economy, and has been a very fulfilling and enriching experience.

I thank Michael (Mike) Smith for making some invaluable connections. First and foremost, he connected me to his outstanding literary agent, Rafe Sagalyn. I am grateful to Rafe for sharing insights about the book publishing process and for navigating discussions with the MIT Press. Mike also connected me to Jane MacDonald, who was my first acquisitions editor. She believed in the thesis of this book and I am grateful to her for that. Midway through the process, the wonderful Emily Taber started to handle my manuscript and became my acquisitions editor. I am deeply grateful to Emily for being the beacon that eventually lit my path to publication. She guided me through the publishing process, shared invaluable insights throughout the process, and provided excellent editorial comments.

I thank the following folks for giving excellent feedback that helped improve the manuscript: Panos Adamopoulos, Amitabh Bose, Gordon Burtch, Beibei Li, Martin Spann, Vilma Todri, and David Verchere.

I am thankful to Jessica Neville of Neville Communications and to Frank Luby and Elana Duffy of Present Tense for editorial support and feedback.

Over the years, the NYU Stern Public Relations team has been a terrific source of support in taking my research and helping to disseminate it widely. I would like to thank Jenny Owen, Beth Murray, Rika Nazem, Jessica Neville, and Janine Savarese Lowe. I also thank Janine for introducing me to the book publicist Rimjhim Dey. I thank my wonderful past and current colleagues in the NYU Stern administration who have helped me in numerous ways over the years: Joy Dellapina, Naomi Diamont, Megan Hallisay, Sharon Kim, Shirley Lau, Mandy Osborne, Laura Shanley, Ashley Tyrell, and Alanna Valdez.

I thank my parents for always believing in me and for their unwavering support. I thank my in-laws for being a second set of parents for me. My sister, my two brothers-in-law, and my sister-in-law have been an incredible source of help and support.

My world changed completely and in the most spectacularly delightful way the day our daughter Ananya was born. Her presence in my life motivates me to work harder and continually strive to improve myself. She has taught me the true meaning of unconditional love. I hope that, when she grows up, she will be proud of me for penning this book.

Last but not the least, I am deeply indebted to my wife, Deepti Shrivastava. Not only did she painstakingly undertake several round of high-quality edits to my book, but as a seasoned brand marketer she also helped refine many of my thoughts. I thank her for letting me chase my dreams and climb mountains, both literal and metaphorical.

Introduction

Imagine a future in which parents have their child wear a wristband when the family is at the beach, and a mobile app beeps to alert the parent if the child strays outside an area defined by the parents. Or a future in which someone could travel to a city for the first time and immediately find not only the kind of cuisine, clubs, bars, and stores he likes, but also receive personalized offers that lure him in. Or a future in which a company could send a coupon to a potential customer before she even leaves for a shopping trip she didn't even know she was going to take. Or a future in which, instead of flipping through a physical catalog and guessing how something might look in your home, you can go onto an interactive platform and place products in your rooms, using an app and the cameras in your mobile device.

That future is now. In fact, it's so now it is almost yesterday already. None of these examples is science fiction. Companies are creating and improving this reality as you read this.[1]

The technological ability already exists to make our lives more efficient and fun. Its biggest enabler is right in the palm of our hands and often in our pockets. It is the smartphone. It is no longer just a device. In 2015, the mobile ecosystem generated 4.2 percent of global GDP, a contribution that amounts to more than $3.1 trillion of economic value added.[2] The former CEO of eBay, John Donahue, referred to mobile devices as the "central control system of consumers' lives."[3] The always-on lifestyle has become so pervasive that we now take it for granted. But many will tell you that being emotionally and spiritually attached to our phones is a choice we all make. Is it?

Motivation Behind This Book

What made me think about writing this book? I believe it was in late 2009 or early 2010 that I first heard the phrase "we live in an era of smart phones and stupid people." I was very amused with this statement but I was also intellectually intrigued. Not so much about what smartphones would eventually do to human intelligence, but more by how it could become an exceptional source of intelligence for businesses. My instincts were telling me that the smartphone would have far-reaching and profound implications. But I craved concrete evidence. Could the crystal screen of the mobile phone have the potential to become a crystal ball for businesses? I felt compelled to answer this question.

It is this intellectual curiosity that led me to adopt a multi-year, multi-country research agenda on understanding the potential of the mobile economy while at NYU Stern School of Business and, for a short while, at the Wharton School of the University of Pennsylvania. During my days as a doctoral student at Carnegie Mellon University, I had become fascinated by how the Internet and other digital technologies were rapidly transforming markets and industries. Between 2004 and 2009 that curiosity led me, as a young researcher, to embark on a journey that would take me into the world of Internet commerce and then deep into the world of social and digital media. And then one fine day I came across the above-mentioned quotation about smartphones. It occurred to me that the mobile phone was going to be perhaps the most profound and transformative source of impact on business and society we had seen.

Between 2009 and 2016, research projects, industry consulting assignments, and other engagements on the topic of the mobile economy took me to many countries—among them Brazil, Canada, China, France, Germany, India, Italy, the Netherlands, South Korea, and the United Kingdom—in a quest to understand what consumers do with their smartphones and how businesses can use that understanding to improve their products and services.

I was struck by how similar consumers are across the world when it comes to their mobile phone usage behavior. Between the West and the East, between North America, Europe, and Asia, I expected major differences in consumers' interactions with brands on mobile devices. These regions are culturally very different, after all. But I found very little variation. Smartphones have rapidly, universally and permanently created a shift in people's wants and expectations from businesses. *People are willing to exchange their information with businesses in exchange for relevant offers that generate concrete value for them. But it has to be done in a way that preserves consumer trust in businesses' use of their data.* Therefore, businesses have an excellent opportunity to create value for their consumers, earn trust and build a relationship through their interactions with consumers on the mobile device. This won't be a one-off event. It has to be a journey.

I have benefited greatly from my interactions with my colleagues and other academic thought leaders during my academic research projects. I learned a tremendous amount from my meetings and conversations with senior corporate executives and startup founders across the globe during my corporate consulting projects. While teaching in many different executive education programs around the world, I acquired invaluable nuggets of wisdom from the participants who hailed from five continents. While keynoting conferences, I received indispensable feedback from many domain experts. During these sessions and meetings, I repeatedly encountered similar kinds of questions from the audience that had one common theme: What are the various factors that influence the world of mobile marketing and how are they shaping the mobile economy? How can we unlock mobile marketing's vast potential?

I then realized that there was a great deal of enthusiasm in people in wanting to see a book like this—one in which much of the existing academic work in this space comes together in an integrated manner. It was a non-trivial challenge. A mainstream book based on peer-reviewed academic articles not only needs to encompass the rigor of academic science but also needs to be written in an interesting and accessible way. I was cognizant of the fact that the important insights from our academic studies should not get lost in translation or in technical scientific jargon. Presenting the most surprising, compelling, insightful, and actionable findings from the academic literature of the past 10 years or so with maximum clarity is my goal with this book. And I hope it comes across clearly to you, the reader.

Genesis of the Title *Tap*

Consumers tap into their smartphones, and swipe them, thereby creating a data trail. Businesses can then tap into (that is, draw upon) this trail to predict our preferences and curate offers for us. This double entendre is the origin of the book's title. This two-way street creates a feeling of intimacy and connection. I am sure some of you have had an uplifting feeling when a marketing offer popped up out of nowhere at the right time and helped you. We attribute it to coincidence, karma, luck, or fate. Now imagine if those feelings occurred more frequently and reliably, but it wasn't coincidence, karma, or luck. It was all planned in advance, driven by data, curated just for you. That is the world I describe in this book, a world we should feel comfortable with and start getting excited about.

Today consumers get exposed to way too many irrelevant and redundant messages from firms. This happens because firms often do not have the kinds of data about consumers' preferences to generate the most pertinent offers. As a result of this lack of information, consumers in the digital world get bombarded with irrelevant ads and offers. From the firm perspective, it is akin to throwing darts in the air, hoping one of them will hit the bull's eye (which, in this case, is to draw some form of consumer engagement from the ad exposure). In turn, the high frequency and low relevancy of these messages make consumers even more annoyed.

The good news is that the vast potential of the mobile economy can change this vicious cycle. What this book will show is that whenever brands have been able to access and harness data on consumer preferences from mobile and digital devices, they have been able to curate highly germane and non-superfluous offers. Such data-driven evidence makes me confident that we can create an exciting reality that is a win-win for both firms and consumers.

What This Book Offers You

For anyone responsible for marketing, advertising, media, or making data-driven business decisions, the entire process of harnessing the power of mobile data and delivering value to consumers can be intimidating. One of my objectives with this book is to demystify the mobile economy and take the uncertainty out of that process. The insights in this book, drawn from groundbreaking research, cutting edge case studies and inspired experimentation, will help brand and media executives, marketers, advertisers, technology professionals, analysts, current and future entrepreneurs, business and economics students, data scientists, cross-functional executives, and policy officials understand the mobile economy. It will help them harness the power stemming from many different facets of the mobile economy in ways that benefit their companies, their organizations, and their customers. By reading *Tap,* they will learn the psychological mechanisms that shape peoples' mobile phone usage behavior. The reader will learn how to strike the right balance as they and their target customers grow accustomed to the connections and interrelationships that flourish when firms start tapping their potential. They will learn about the nine forces that can be leveraged, both individually and in combination, not only to transform the mobile channel into a highly effective medium for increasing engagement and customer satisfaction but also for improving their revenues and profits.

If you are responsible for branding, marketing, media, or advertising decisions, *Tap* will show you why it is time to seize the opportunities created by the global prevalence and dependence on smartphones. The insights in this book will empower you to tap into unprecedented opportunities that are headed our way in the mobile economy, resolve several behavioral contradictions displayed by consumers, and delight consumers all at the same time. *Tap* will demonstrate the true power behind mobile marketing: the influence it wields over shoppers, the behavioral and economic motivations behind that influence, the lucrative opportunities it represents, and how you can start to rethink your marketing strategy on the basis of these insights. Whether you are marketing products or services in telecom, retail, banking, insurance, hospitality, ecommerce, health care, manufacturing, or really any other field, *Tap* will expand your current understanding of how the mobile ecosystem is shaping the digital economy, how to participate in this revolution and what the future will look like. You will come away with new concepts that give you and your firm an edge by maximizing the impact of your mobile strategies. It will help you get started and engage in a meaningful conversation with domain experts.

Finally, *Tap* is fundamentally a book about human behavior. If you like to understand how social psychology intersects with business economics and in the mobile economy, this book is for you. If you are inspired by the latest technological innovation that enables businesses to tap into the world of opportunity that is our mobile phones, this book is also for you.

Your Smartphone: Your Personal Concierge

Today, many businesses can get to know their customers before they walk in through the door simply by identifying patterns in their data and making connections between the dots. The amount of data consumers transmit through their mobile devices is so extensive and so rich that marketers today can do things their 20th-century counterparts could not even fathom, and marketers a few years ago could only dream of. Yet even now, nearly 20 years into the 21st century, marketers have only just begun to tap into this opportunity to reach consumers directly, make immediate, on-the-spot offers they can't resist, and engage them in ways that make digital advertising seem more like a helpful service. They are yet to transform the smartphone into our personal concierge, our butler. And while marketers have yet to harness or even understand the power of mobile marketing to its fullest extent, we can make this a reality we can all share right now.

Mobile marketing is a powerful approach that other forms of marketing and advertising—from direct to TV to print to pop-ups and even search engines—cannot come close to matching. Mobile has most of the advantages of those advertising forms, few of the disadvantages, and also has features and capabilities of its own. The reality is that in the digital world most people find advertising annoying, overwhelming, or intrusive. They dislike ads that ruin their browsing or consumption experience and strongly dislike ads that are irrelevant or redundant. Hence, I cannot stress enough how big an opportunity it is for businesses to rectify this problem. Done elegantly, mobile phones can make advertising seem like a valuable service consumers want to tune into, and less like a disruption or nuisance that consumers try to tune out. Businesses need to build a future in which a mobile device can become the personal concierge of their consumers.

Building This Future

This kind of a future starts with data, of course. And the more consumers empower their smartphones and inform it, the better it works for them. Making that future a reality will also require firms to be imaginative, creative, and transparent in how they seek out data from consumers. One of the best examples of this is how Facebook allows users to give them feedback on what ads they find relevant. This tool, known as Ad Preferences, is accessible from every ad on Facebook that explains why any consumer is seeing a specific ad.[4] More importantly, it lets users add or remove preferences that Facebook uses to show them specific ads. In essence, it lets people have more control over the ads they see.

Building this future will require tradeoffs, some of which consumers make reluctantly. Consumers let businesses have and use their data, as long as they provide something of value in return. Figure I.1 shows the types of intimate data consumers could share with firms, with the expectation that the firms will use it to make their lives easier and more fun. And with this intimacy often comes vulnerability. Of course, consumers have to be somewhat careful as they proceed with this intimacy. The increasing volume of data naturally puts personal privacy at greater risk. Not everything that enters through the smartphone's two-way portal is beneficial and benign. These devices store everything from credit cards to banking information to data on travel patterns and other sensitive information such as home address. Businesses increasingly see consumer data as an asset, and rightfully so, but it can become a liability if the data end up in the wrong hands.

Figure I.1
The kinds of data consumers can share with brands and retailers. (adapted from Phil Hendrix, The Engagement Stack, "a whitepaper sponsored by Brandify")

Put the phrase "data breach" into a search engine and the results read like a "who's who" of major merchants and service providers.

The operative word is "balance." The mobile channel is more immediate, holds more value, and more potential for firms trying to reach a connected market of smartphone users. To earn the right to build and nurture close personal relationships with customers, companies need to remain vigilant about maintaining a balance across gathering data, using the data for the mutual benefit of the customer and the company, and protecting the data from unauthorized eyes. In the same vein, companies need to be careful about asking too much of customers before they are in a position to create value from that information.

This form of marketing is still new. Companies can't risk breaking relationships and damaging trust. What should be heartening for businesses is that an increasing number of users do not object to ads *per se*. What they often object to is a brand that is intrusive, or always trying to do an overt hard sell, or both.[5]

How can companies mine users' behavioral patterns to surprise and impress consumers? How can they send only advertisements that make consumers' lives more efficient and cut their search costs? How can they give users relevant choices instead of maximum choices? How can they collect and use more and more data from consumers but still reassure them that they are safeguarding personal data?

I argue in this book that companies need the ongoing cooperation of consumers. But they also need the breadth and depth of insights—plus imagination and creativity—to generate the most value from their mobile marketing efforts.

Outline of Chapters

Part I

The chapters in part I of the book lay out a compelling case for how the mobile phone has enabled consumers to see and understand many facets of human behavior: how we interact, how we shop, what our habits are, and how companies can use these insights and data to their customers' advantage and to their own. They also lay the groundwork for later analyses.

Getting comfortable with and excited about the future I describe may require some self-reflection. For better or worse, we humans have some hard-wired quirks that can make us behave irrationally and get caught up in some apparent behavioral contradictions. From the perspective of mobile advertising, there are four basic contradictions between what we want (or think we want) and how we behave:

1. People seek spontaneity, but they are predictable and they value certainty.
2. People find advertising annoying, but they fear missing out.
3. People want choice and freedom, but they easily get overwhelmed.
4. People protect their privacy, but they increasingly use their personal data as currency.

To relieve the stress and frustration of these contradictory forces, we need to find a happy medium, a sort of balance. The challenge for businesses is to find the right balance between leaving consumers in control and serving them the optimal amount of information they need in order to make decisions. There are nine forces that come into play when determining how best to provide information to influence consumers' decisions: context, location, time, saliency, crowdedness, weather, historical shopping patterns ("trajectory"), social dynamics, and tech mix. Keeping all these forces in mind sounds like a very tall order. But the increasing sophistication of the mobile ecosystem, and the connective tissue of technology that underlies that ecosystem, will make it feasible for businesses to deliver perceived and real value to the consumer in an ongoing and on-demand fashion.

Part II

The phrase "Attention K-Mart shoppers" entered the American popular vernacular in the 1990s, when the retailer K-Mart began to offer "blue light specials"—a marketing tactic designed to draw shoppers toward an instant bargain somewhere in a store. An employee would wheel the blue light—a police-car-style light mounted on a pole higher than the store's shelves—into position next to the discounted products, turn it on, then make an announcement over the store's public address system. The blue light would compel customers in the store to make short-term decisions (consciously or not) by mentally answering a series of questions: How far away is the light? What is on sale? Do I need it, and do I need it now? Am I in a hurry? Am I short on or flush with cash? Why am I even in K-Mart today? How much is the discount?

Well into the 21st century, the very same questions still circulate in customers' minds when they are confronted with an incentive to buy something. The same processes go on for a text message as for the blue light in the linen aisle at a suburban K-Mart. The difference is that today marketers don't need to rely solely on gut feeling and experience and turn on a blue light. Marketers now have data—lots and lots of data—and can use it to guide consumers' decision making. The chapters in part II go deeper into the nine forces that drive consumers' purchase decisions: context, location, time, saliency, crowdedness, weather, trajectory, social dynamics, and tech mix. They explore how that device in our pocket, the one so many of us can't live without, unifies all of the nine forces I just mentioned into one powerful platform. Though we have had this technology for only a short time, we can already derive some principles, strategies, and tactics that will improve our understanding of the influence these devices wield over shoppers, the behavioral and economic motivations behind that influence, and the monetization opportunities those motivations represent for firms across a wide variety of industries.

It is easy to draw intuitive and correct conclusions about these forces in isolation. Yes, our intuition tells us that a consumer is more likely to visit a closer store than one further away, or to choose a purchase option that appears closer to the top of the results delivered by a search engine or a shopping app. Nicer weather brings out more shoppers, some on foot and some in cars. The power of mobile advertising lies in the *combination* of these forces. The likelihood of redemption increases as the price and the distance decrease. That is intuitive, but now we can turn this kind of intuitive qualitative insight into a *quantitative* one. We can make it measurable and even precise. The magnitude of each force, in isolation, matters. But what matter a lot more are the interactions between them, which truly unlock the power of mobile advertising.

The chapters in part II are organized around the nine powerful forces behind consumers' purchase decisions. Each chapter will take you through the research and analyses that show how companies can confidently and strategically influence one or more forces through new emerging technologies such as geo-fencing, geo-targeting, and geo-conquesting. The studies are based on real consumers' responses in the real world—in subways, in shopping malls, online in ecommerce platforms, and offline in physical stores—rather than in the lab. The collective insights will lead to food for thought and to recommendations on how to translate data into money in ways that few other strategies and tactics can. When one (or more) of those nine forces is (or are) harnessed correctly, mobile marketing does indeed result in higher redemption probability, faster redemption behavior, higher transaction value, higher revenues, higher engagement, higher customer satisfaction, and so on.

No other platform is better suited to capitalize on all nine of these forces than the mobile channel. It combines targeting with immediacy and context in a way that delivers the consumer the right individualized incentive for the right product in the right store in the right place at the right time.

A vast and fast-growing body of academic research has shown that under certain circumstances the counter-intuitive exceptions to these "rules" can throttle the very impulses advertisers are trying to stimulate. Some behaviors that hold true during the week work less well on weekends. An advertising approach that seems obvious for a certain combination of location and time of day may be less successful or even ill-advised if the target customer's social dynamics change or if the firm is unsure of the consumer's shopping occasion.

With focused investments, companies will not only transform the way consumers view and interact with their products and brands, but will also find the right customers more easily, target them better, convert them faster, reward them sooner, and keep them longer. And companies are creating and improving this reality as you read this.[6] But the business side of this equation isn't the only winner.

When companies challenge themselves and put their insights and data to creative uses, consumers can find more relevant products, work with the providers who best match their needs, manage their short-term and long-term "to do" lists, plan ahead, and make better and more informed decisions. And they can do all this faster, more efficiently, and on a much larger scale than they could have imagined even 5 years ago.

This is generating an openness to advertising, especially on something as personal and close to us as our hand-held devices. More and more consumers, especially those under the age of 40, are warming up to the notion that there is a give-and- take relationship between them and the businesses that hope to serve them.[7] Sure enough, there are some geographic differences. Consumers in Brazil or China are more likely than those in the United States or the United Kingdom to click on a relevant mobile ad.[8] But even in the United States, consumers' behavior is changing rapidly as they become more amenable to interacting with mobile ads. The studies and experiments discussed in this book are testimony to the fact that consumers around the world are willing to take small but steady steps toward developing trust with marketers, as long as the marketers are willing to move toward the balance mentioned above and to provide them real value in return. Give-and-take is what defines these digital relationships and makes them work well for both parties.

The mobile phone is an excellent medium for marketing that should maximize the benefits to consumers and minimize the intrusiveness. Consumers have made it clear that if advertisers engage them appropriately on mobile devices it can have a huge impact.[9] Businesses can transform our smartphones to act as our personal concierges—our butlers—and not as stalkers.

Part III

Part III offers my view of the future. In it, I predict that the integration of mobile devices with other smart devices will have a dramatic effect on the world of business. The other devices include wearable technologies, artificial intelligence, messaging apps, chatbots, deep linking, smart televisions, connected cars, mobile payments, and virtual and augmented reality.

Tap gives the reader an inspirational and aspirational view of what is to come for the mobile ecosystem not only in the world of business but also in society as a whole. In the coming years, firms will be able to predict consumers' behaviors better, and to increase the amount of fulfillment and convenience in our lifestyles by providing better and easier choices. At the same time, some concerns about data privacy are certain to be raised—concerns to which both firms and consumers should pay careful attention. Firms can alleviate these informational privacy concerns by practicing the time-tested principles of *notice* and *choice* by notifying consumers how their data is being used and by giving consumers a choice thereafter of acting upon it or not.[10]

Mobile phones have the potential to change not only how we shop, but also how we work and interact with others. In the epilogue I discuss how the technological progression will change the nature of work in the world. The same technology that helps us track down what we want and why we want it, and pushes us information when we subconsciously need it, can help us do much more. In the epilogue I also provide a small glimpse into how mobile technologies provide social benefits. In countries where large portions of the population don't have access to bank accounts, mobile phones provide a means of conducting cheaper and safer digital transactions. In rural communities, mobile phones are helping farmers make more informed decisions about what to plant, when to harvest, and how to price their yield. Health workers in rural areas of sub-Saharan Africa use mobile phones to get help from medical specialists.[11]

The future we can imagine and reach in our lifetimes isn't a series of science-fiction thrills such as flying cars and weekend trips to Mars. Of course those are wonderful visions, and they will change our lives dramatically if and when they happen. In the meantime, in the next few years, we can build a better future by fixing and improving a lot of things that may seem more mundane but can enhance our quality of life. We will build this future through the decisions we make each day and the interactions we have with one another—from what to buy and where to buy it to whether to donate money or volunteer with a local charity.

Mobile phones remove a lot of the friction and the frustrations that force us to accept bad compromises. These decisions and interactions may be as complex as finding a new job, as urgent as allocating essential services when a natural disaster strikes, or as simple as purchasing an umbrella at a nearby store, on sale, ahead of a storm, when our device notifies us. No matter what the situation, our lives can improve drastically when we understand and harness the untapped power of that concierge or butler in our pockets.

Whether you have noticed it or not, that future has already started.

I Human Behavior and the Mobile Phone Journey So Far

1 Mobile Phones: A Truly Transformative Technology

To say that smartphones have become a welcome disruption to our daily lives is an understatement. Consumers' adoption of 3G and 4G technologies has outpaced all other technologies, with the Earth becoming home to 3 billion connections in the first 15 years and projected to grow to 8 billion smartphone users by 2020.[1] The Swedish telecom giant Ericsson was even more optimistic in its 2016 mobility report, predicting that the number of mobile phone subscriptions will reaching 9 billion by 2021, exceeding the number of people on the planet.[2] Ericsson anticipates the uptake of 5G services will outpace the uptake of 4G. Led by South Korea, Japan, China, and the United States, 5G subscriptions should reach 150 million by 2021. The 5G wave is the forefront of an ongoing technological tsunami that will see LTE subscriptions reach 4.3 billion by the end of 2021, up from 1 billion in 2015. The Middle East and Africa will also see a progression from 2G to a market in which about 80 percent of subscriptions will be 3G or 4G.

What has trigged this proliferation?

Technological advancements are lowering adoption cost and enhancing user experience. The global average cost of mobile subscriptions relative to maximum data speed has decreased 99 percent, decreasing about 40 percent per year between 2005 and 2013. For mainstream users, smartphones have become much more affordable, approximately 30 percent of all units sold costing less than $100 and some as little as $40.[3]

Though the above advancements have certainly facilitated the adoption of mobile devices, it really is the tremendous social and economic value generated by the mobile ecosystem that has led to the rapid proliferation. A report by the Boston Consulting Group (BCG) shows that mobile technologies are a critical driver of the world economy, generating global revenue of almost $3.3 trillion and 11 million new jobs.[4] Some of the countries reaping the greatest rewards from the mobile economy are the United States, China, South Korea, India, Brazil, and Germany. The BCG study found that in those six countries the combined mobile GDP (mGDP) contributes more than $1.2 trillion to overall GDP (see figure 1.1). This is noteworthy because those six countries account for 47 percent of global GDP.

Naturally, within and across countries, many businesses have benefited too. The mobile ecosystem has been a driving force in the success of some of the world's well-known companies. Six of the 25 most valuable companies in the world—Apple, Google, China Mobile, Alibaba, Facebook, and Verizon—are major participants in the mobile value chain.[5] In his keynote speech at the 2010 Mobile World Congress, Eric Schmidt announced that Google would first focus on mobile devices and all other devices would come second.[6] That statement arguably led to the genesis of "mobile first" attitude in Silicon Valley.

But it not only giant firms that have reaped the benefits. The BCG report also shows that 25 percent of small and medium enterprises (SMEs) that use

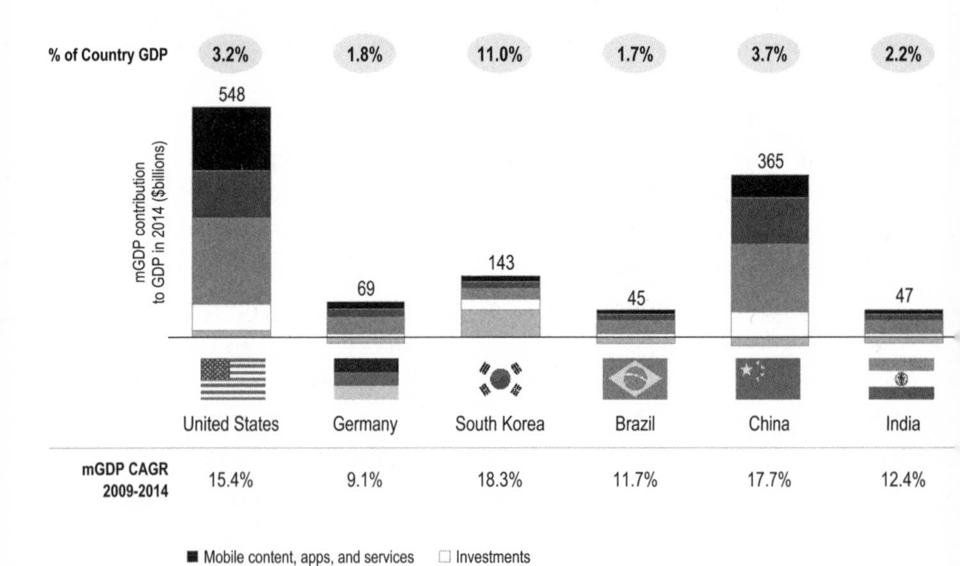

Figure 1.1
Mobile GDP captures about $1.2 trillion in the top six countries by mobile spending. (adapted from Boston Consulting Group analysis)

mobile services more intensively have seen their revenues increase twice as fast and added jobs up to eight times faster than their peers. The mobile ecosystem is also a driving force for fostering innovation in the startup community. According to the BCG report, 7.9 percent ($37 billion) of all venture capital funding in 2014 was invested in mobile economy startups, up from 3.8 percent in 2010. Figure 1.2 captures some of these facts that are a testimony to the impact of mobile devices in this economy.

Last but not least, mobile phones have fundamentally altered consumers' behavior. Whether it is for entertainment or information gathering purposes, we are spending less time on traditional channels such as print, radio, television, and even the desktop Internet, and more time on mobile devices. On average, US consumers spend 10 percent of their time every day looking at a small screen in their hands. Time spent on mobile apps is now exceeding time spent on TV: the average US consumer is now spending 198 minutes per day on apps and 168 minutes on TV.[7] Indonesians have the highest level of active screen time of any country in the world: they consume a daily average of 9 hours of screen media across their digital devices.[8] Few technologies in the history of mankind have proliferated faster and have had a greater transformative effect than the smartphone. And the

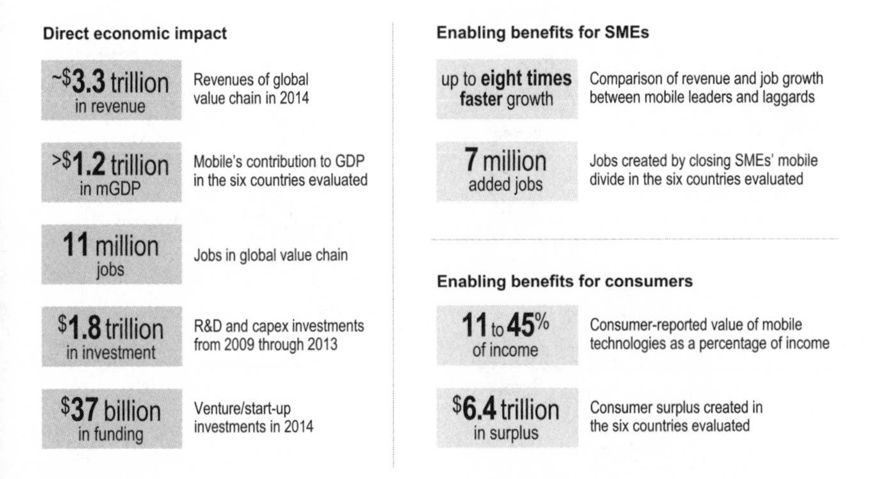

Figure 1.2
Mobile technologies have been a growth engine with a direct economic impact, enabling benefits from SMEs and consumers. (adapted from Boston Consulting Group analysis)

technology continuously improves with smaller parts and faster processors, allowing a continuous flow of new designs and new features to hit the market and render the hit product of a few years ago obsolete.

Despite the increases in time spent on mobile devices over the years, businesses have not kept up with the opportunities for monetization. Figure 1.3 shows how these shifts in media consumption create a huge market potential for firms in the mobile economy, especially in the domain of mobile advertising. The renowned securities analyst and venture capitalist Mary Meeker has been tracking such data for several years. Every year I wait for her "Internet Trends" report with almost childlike anticipation. As of 2015, in the United States, the share of spending on mobile ads (12 percent) is proportionally low relative to time spent (25 percent), which translates into a $22 billion opportunity for mobile advertisers. In contrast, print still

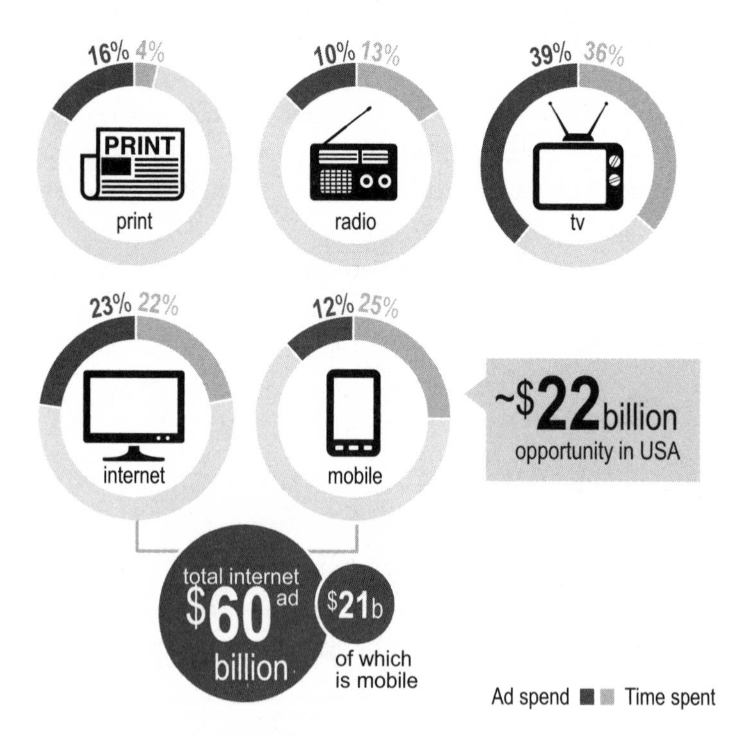

Figure 1.3
Percentage of time spent in media vs. percentage of advertising spending (United States, 2015). (adapted from Mary Meeker's talk at Kleiner Perkins Caufield & Byers Internet Trends Conference 2016)

attracts a share of ad spending (16 percent) that is far out of line with how much time consumers spend with it (4 percent). Of course, I will be remiss if I don't mention that, as more advertising money gets pumped into the mobile economy, one has to keep in mind that the increase prevalence of ad fraud in mobile devices. A recent study from the Interactive Advertising Bureau estimates that mobile advertising fraud cost firms nearly $1.25 billion in 2015.[9] Nevertheless, there is no doubt that the monetization potential of mobile marketing and mobile advertising is huge.

What does this mean for the mobile marketing economy? It means that in this book I am still describing this technology's childhood, if not its infancy. It should also create a sense of urgency. Companies that don't catch up now or cement their advantages now will see their disadvantages grow geometrically as new "killer apps" emerge to capitalize on the new bandwidth.

Looking back to the turn of the millennium, the smartphone now seems like a huge leap forward. In reality we are still at the start of this breathtaking transformation. To put that in context, I would like to take a step back and show you how far and how fast this technology has moved forward in the past 25 years. Project that same momentum out just 5 or 10 more years and you can get a sense of how our amazing opportunities to communicate and interact through mobile devices will continue to grow faster than we can seize them.

Evolution of the Mobile Phone

We haven't always had it this good. A smartphone, according to Merriam-Webster, is "a cell phone that includes additional software functions (as e-mail or an Internet browser)."[10] By that definition, the smartphone did not even exist until 1992, when IBM introduced the Simon Personal Communicator. For about $900, the Simon would let you send emails and send faxes from a touchscreen component. It had a base station and ran approximately 60 minutes after a full charge.

IBM sold approximately 50,000 Simons in 6 months before pulling the product from the market.[11] By comparison, Apple sold 50,000 iPhones *every 90 minutes* in the fourth quarter of 2014.[12] It wasn't the hefty price tag or the fact that the Simon weighed more than a pound that deterred mass consumption. This was average for the most modern technology of the early to mid 1990s. What killed the Simon was the genius of invention. The innovation was simply too far ahead of its time. It was launched before email was popular. Phone networks were not structured for data, nor could they keep up with the demand. The usefulness of "apps" in this primitive form was limited. Having the optional camera was a nice idea, but without software to make it possible to do something with an image after capturing it, its value was limited.

"Convergence" became a buzzword as we approached the year 2000. How quickly we forget that in those days not everyone possessed a home computer or even a basic cell phone. But many people carried around cameras, portable game devices, portable music players, and a paper calendar of some sort. "Notebook" meant something spiral-bound with lined white paper inside.

By the mid 2000s, personal computers were almost ubiquitous, and cell phones were in nearly every pocket, at least in the more developed nations. Google was a household name. Smartphones existed in various forms, such as the Blackberry with a keyboard option and the Samsung model with a touchscreen and a keyboard. The Palm OS allowed us to achieve some convergence by fitting schedules, calendars, and other tools into our pockets. The time was right for an innovative smartphone that would be dependable an, above all, relevant to a mass market. Enter Apple.

The first iPhone, released in 2007, boasted many of the same features as other phones, with a flashier display and fewer buttons. This made it attractive outside of the business world, where most people relied on their smartphones for their email and scheduling capabilities. Perhaps more importantly, the iPhone operating system allowed for all new Web-based code. This permitted websites to display in a better, more readable format on a phone. The smartphone was now good for much more, from Google search to maps to shopping for shoes while waiting for a bus to social networking on Facebook.

Evolution of Technological Capabilities

Network capabilities have also improved enough to erase the memories of dial-up modems and clogged phone lines. The advent of Wi-Fi technology eliminated those problems. Through its ubiquity—in homes and in many public spaces, such as coffee shops, airplanes, and subway tunnels—Wi-Fi not only created an opportunity for 24/7 connectivity but spawned an *expectation* of it. We can be connected all the time, no matter where we are.

Precision technologies have also changed our lives. No matter where we are, chances are very good that someone can find us if they try. Geocoding, beacons, Wi-Fi, RFID, and near-field communication allow for proximity notifications, marking our presence and helping to determine the greatest benefit we could experience in that immediate area, from meeting up unexpectedly with a friend who is nearby to taking advantage of a sale at a favorite store. The launch of iBeacon by Apple in 2013 triggered a global growth in the proximity solution provider (PSP) industry. There are 293 PSPs as of 2016, with 38 percent headquartered in the United States, 10 percent in the United Kingdom, and 26 percent divided among Australia, France, Germany, Italy, the Netherlands, Norway, and Spain.[13] New PSPs are emerging in India, Vietnam, Singapore, and Latvia. This is not an isolated phenomenon; it is a global force.

Just as Google's indexing of the World Wide Web paved the way for consumers to discover firms online, PSPs will index the physical world to connect it to the advertising industry.[14] But as these PSPs proliferate, it is important for them to be cognizant of the potential risks and downsides. For example, how are location data being shared? Who has access to the data? The majority of the public does not fully understand location data, and the majority of businesses need to know more about the management of location data.[15]

The next generation of these technologies includes mobile wallets, which will virtually eliminate plastic and paper from day-to-day commerce. Using mobile wallet services, consumers will be able to manage and use offers, coupons, loyalty rewards, tickets, boarding passes, gift cards, identity cards, electronic receipts, and product information from multiple brands in digital form. Scan a bar code with your smartphone and you get a wealth of product information immediately. While you are reviewing that information, an app will search online for the same code and compare your current deal for you. You can then buy the item right then and there, using a card stored on our phone, even if you are standing in the aisle at a competitor's store. On-the-go magnetic stripe readers offer similar advantages. Apps such as Square can process payments from a portable card reader. This has helped make small businesses and individual sellers more competitive. It has also helped large businesses to enhance consumers' experiences in stores by allowing floor attendants to check out customers on the spot instead of forcing them to wait in line at the cash register. A modern, attractive ecosystem has blossomed from these technologies, and it is poised to thrive as mobile marketing becomes central—if not essential—to consumers in many parts of the world.

The world opens up for us as consumers when we have the Internet in the palm of our hand. At the same time, we open up ourselves to the world. When we spend the better part of our waking hours tapping into our screens as part of our daily lives, our activity leaves a rich and tantalizing trail of data. Our smartphones and the apps on them collect and transmit most of the data without our conscious, direct intervention. This step change in data availability and collection is further enabling businesses to tap into data about consumers and to shape their lives. But at the same time businesses have to be extremely mindful of how they manage and protect the data. Location data privacy management is challenging because location data are becoming more complex and is uniquely sensitive because it acts as a common denominator linking multiple data sets.[16]

Sometimes people are casual about their data because they are either unaware that their data are being captured or unaware of the scale at which location data are being collected. As some observers say, "at best users may get an innocuous 'this app would like to use your location' alert, which masks a lot of what is really taking place and what that ultimately means from a personal privacy perspective."[17]

2 What the Smartphone Has Changed

The last 15 years have seen rapid evolution in mobile phones and related technologies, but the most noteworthy change has been their impact on consumers' behavior. Consumers around the world encounter modern mobile technologies so often that they take them for granted. Their ubiquity has led us to overlook the *collective* impact of these technologies, how pervasive they are, and how they shape even the most mundane of our daily activities.

Take a typical day in the life of the regular person. When we wake up in the morning, we tap the crystal screen to check our email and any instant messages. We check our favorite social network sites for updates from friends and family. We check for breaking news from the usual websites or apps. We check our calendar to remind ourselves what the day looks like. Then we check the weather to see how we should dress for the day. We open up maps to check commute times and the fastest route to work. Our morning routines have been forever changed or enhanced as a result of smartphones.

For more on how consumers' behavior has changed, take a closer look at communication. We now have the ability to reach out and touch someone instantly with a quick message on WhatsApp, or to know when a friend is nearby (via Facebook's Nearby Friend tool) so that we can set up an impromptu meeting. Domestic long-distance phone call plans are history, which means we no longer need to consider location when establishing and budgeting for communication. Even phone calls themselves are losing their primacy as a means of communication. Voice minutes among those in the 18–34 age group fell to 900 minutes per month from 1,200 in 2 years after the launch of the iPhone. Texting rose from 600 messages per month to 1,400 messages per month during the same period.[1] Phone calls are increasingly seen as intrusive. They force us to pay attention, to listen and speak in turn, and may even seem menacing to some. Texting is faster and easier, and it leaves us able to multi-task while we wait for a formulated response, or come up with one of our own. It has now become socially acceptable to send an instant message on Facebook, Twitter, or WhatsApp, depending on whom we are trying to communicate with. Improvements in the feature itself have certainly played an important role as well. Before the advent of the smartphone, texting was hard work. Just to begin to write "Come over later," you had to tap the "2" key three times for a C, then tap the "6" key three times for an O, then wait a moment, then tap the "6" key once for an M, and then tap the "3" key twice for an E. These changes have had drastic effects on how we talk to each other. With the right insights, imagination, and creativity, they will also have a drastic effect on how businesses inform and influence customers.

This always-on device has also started to alter our overall social and professional interactions. The 24/7 accessibility has blurred the definitions of work and leisure time, as people surf and take care of personal tasks during working hours, and get work done on what used to be their demarcated personal time. This puts the onus on the employee to set time boundaries and reset priorities. One advantage is that the universal connectivity enables us to turn formerly unproductive time such as traffic jams and flight delays into productive time. During our commute from home to work or vice versa, we can send or receive emails, get caught up on business memos, access digital files from the cloud, engage in business transactions with colleagues on the phone, schedule meetings on our calendar, and prepare for the work day ahead. We can now have multi-way business conference calls while dinner is being fixed at home or work from home while caring for a sick child. These newfound efficiencies can then allow some people to schedule appointments and offline work more efficiently and ultimately spend fewer hours in the office. Smartphones also allow us to have face-to-face video communications with anyone around the globe, thereby enabling a "work from home" culture. Not everyone perceives these changes as advantages, though. More than half of employers complain that smartphone use, including texting, is a big drag on productivity.[2] In addition, another complaint is that smartphones create an expectation of "always on" and the intrusion of work into personal space. Despite these criticisms, there is no question that telecommuting and using downtime productively is an opportunity for managing work flows better. The smartphone makes that possible.

There are many other ways in which this on-the-go device has shaped our behavior. For instance, access to information has never been greater or more customized to suit specific needs, thanks to mobile devices. We can make airline or hotel bookings on travel apps. Whether it is searching for the nearest restaurant on Yelp, finding a short-term bike rental on Citibike, or sharing rides at very short notice on Uber or Lyft, smartphones have enabled this kind of on-the-go matching at a scale that is unprecedented. These services make use of technological advances such as GPS on our smartphones. Thanks to mobile apps, brands in many industries have lower barriers to entry and can scale up quickly. Take the sharing economy (or the collaborative economy, as some experts call it), for example. Perhaps the most profound enabler of the sharing economy has been mobile apps. In the absence of smartphones, it is anybody's guess if these innovative solutions would have seen the light of the day.

The proliferation of smartphones has changed how we create and consume any kind of content. Whether consuming streaming music on Apple Music or Spotify, uploading pictures and videos on a social networking site such as Snapchat, Facebook, Twitter, or Instagram, writing a review on Amazon, or writing a comment in response to a favorite food blogger's latest post, we have the ability to create or consume content in a frictionless manner.

How we shop now is fundamentally different from how we shopped ten or even 5 years ago. Back then when Internet shopping first became prevalent consumers used to browse online but make the actual purchase offline. Today, thanks to mobile devices, many consumers often browse in a brick-and-mortar shop and then complete the transaction online on their devices. We can routinely price check, check inventory or comparison shop on Amazon while we are physically inside a store. Shopping is basically on-demand now, thanks to the mobile phone. Perhaps one of the biggest changes (and more relevant for this book) has been in how businesses interact with their audiences. Let's take a closer look at the world of mobile advertising.

Think of someone walking down the street with a copy of a print magazine such as *Time* or *Cosmopolitan*. No matter where he or she goes, the consumer takes the same physical copy of the magazine along. It never changes. Now freeze the frame as the consumer is passing a store that happens to have placed an ad in the magazine the person is carrying. This snapshot, this moment in time, is lined up perfectly to influence the person to visit the store.

Now imagine that you, as the advertiser, can alter the advertising copy in the magazine—who advertises, what product, what message, what format—depending on what part of a city or even which block of a city the reader is in. While this is obviously inconceivable in a print magazine, the smartphone allows advertisers to take such snapshots and target their digital advertisements to a particular segment of readers. This is advertising based on location: where a target consumer is at any given time, accompanied by additional data (such as time of day) to put that location into context. This kind of marketing falls under the umbrella of location-based services (LBS).[3]

LBS has truly become a global phenomenon. For example, Stores Specialists Inc., the largest franchiser of lifestyle brands in the Philippines, launched a location-aware mobile app for luxury-brand shoppers that offers promotions as shoppers approach any of its 500 stores. In Singapore, as customers approach the 313 Somerset mall, they get exclusive promotions and offers from the 170 stores inside the mall via the mall's app.

Around 90 percent of the estimated 153 million smartphone owners in the United States use location services on their phone, according to data from the Pew Research Center.[4] Consumers still have some personal control, because they can switch the feature off and on. Locations can also be partitioned, meaning they are defined or protected by virtual barriers or fences. This helps ensure that the information exchanged is pertinent to both the sender and the recipient. That ability alone represents a huge paradigm shift in how organizations can reach their audiences and interact with them.

The power of LBS extends well beyond marketing. It has brought about changes in society. Cities and other organizations use it for traffic management, emergency alerts, and other critical real-time information. When a fire in San Francisco posed a threat to surrounding buildings and caused extensive traffic congestion in the immediate area, the city used LBS to send mobile notification via AlertSF to people in the affected areas.[5] Uber and Lyft use similar LBS technologies as they begin operations at Los Angeles

International and at other airports that were once reserved for traditional taxis. In order to reduce congestion, Lyft and Uber have implemented geofences, which confine all requests to a dedicated lot where the drivers have to wait. The app must route all requests only to drivers in the correct location.[6] This benefits all parties: the companies, the airport, the users, even the drivers.

Collected data become richer and more useful when we combine location with social media. Some city governments search for certain words in posts on restaurant search apps to zero in on potential violations of the health code and send inspectors out. Cities are capturing such data as quickly as marketers are. Some city governments scan for the word "pothole" in geo-tagged social media posts in order to find the worst, most traveled areas and fix them.

The city of Boston works with Street Bump, a startup that runs vibration analysis in the background of a driver's or a passenger's phone. The analysis identifies road condition and damage and informs the city for repair.[7] Anyone who has driven on cobblestone streets missing a few stones after a Nor'easter snowstorm can appreciate how location can be a huge asset to community improvement, by enabling more efficient and timely delivery of services. All this is possible because people let apps run in the background on their phones.

All these interactions with mobile phones are creating highly specific data about consumers. In 2016 the sheer volume of data available, the granularity of the data, and the accessibility of the data is mind-boggling. Enlightened and empowered by this ability to understand consumers at this "sub-atomic" level, marketers can now devise strategies and tactics that serve consumers rather than exploit them or irritate them. Remember that I mentioned a little while back that firms can use mobile as a concierge and a butler, not as a stalker? Showing the right ad to the right consumer at the right time at the right place on the right device is no longer the realm of Hollywood movies such as *Minority Report*. It is very much a reality today. The key, of course, lies in understanding the different drivers of customers' shopping preferences and how they work both separately and together to influence consumers' decisions. But it looks very rudimentary compared with what advertisers can do in 2016 and what they will be able to do in the future if they collect the right data and analyze them the right way.

One of the most profound changes enabled by the smartphone is the indexing of our movements in the physical world to create our offline profiles. By examining the human trace data recorded through mobile sensors, firms can now better understand the dynamic social interactions of humans and idea flows in society.[8] So what do I mean?

Our tapping on mobile devices is creating a digital trail of data from which firms can formulate a trajectory of our physical movements. This is enabling a paradigm shift that will take marketing from location-based services to marketing based on an individual consumer's *trajectory*. For those who have read the Harry Potter books or seen the films, understanding trajectory is similar to the concept of the Marauder's Map, which shows the location of every occupant in a building and footprints where that person recently walked. Classic Looney Toons cartoons often used the same idea, showing the footprints of two foes as they pursue or fight each other.

Now imagine such a map for an individual consumer. It traces the movements over brief periods, then combines them over time to create more elaborate paths. This is the consumer's trajectory. As long as the individual carries the smartphone, it is generating location data either from a GPS device, from a Wi-Fi connection, or from an app that is running in the background—data that trace where the individual went, how long he or she stayed, and how long it took him or her to go from place to place, all in the physical world. Considering the significant search costs consumers incur in the offline world, such physical behavioral traces of individuals can be highly informative in uncovering the preferences that influence a consumer's real-time decision making. Hence, marketers will know where consumers have invested time and effort in offline searches, and in many cases can causally infer why they behaved so.

This kind of offline trajectory information is analogous to the data on search queries and on click streams that are available from users' online browsing behavior. Amazon, Alibaba, and other big online retailers have long used data such as where a person has been online, why he or she was online, and what he or she purchased as a means to understand what interesting products to advertise the next time that individual visits their website. Mobile technologies now allow us to digitize similar kinds of individual behavioral trajectories in the physical environment. Using a person's history of travel through a shopping mall or down a main shopping street provides clues about his or her desired shopping behavior. These clues are so reliable that a marketer usually knows what to expect when the consumer follows the same pattern in the future. These clues act as a filter by signaling with greater clarity to marketers when an advertisement has a high chance of success and when an advertisement can be detrimental because it is a mismatch for the consumer's current context.

Of course, tracing trajectories and using them for marketing purposes raises red flags for some consumers. Tracking movements may seem very personal and much more invasive. While it can deepen the intimacy of the relationship between business and consumer, it can also heighten the feeling of vulnerability. We have seen similar resistance from people who oppose automatic toll payment methods (because they track where people have driven) and the use of security cameras (because they consider it to be surreptitious spying). There is good reason for this resistance. Instances in which someone has been arrested on the basis of location data extracted from a mobile phone and the police officer later discovered that he or she made a mistake are not uncommon. In 2002, in Portland, Oregon, Lisa Roberts was wrongfully convicted of having murdered a former lover. The conviction was based on an error in the interpretation of cell phone tower location data from Verizon.[9] She was eventually released in 2014 after spending 12 years in prison.

But a significant portion of the population is beginning to see such tracking as a source of convenience rather than a concern (again, I remind you of the butler vs. stalker analogy). Avoiding long idle times at tollbooths and enjoying the comfort and potential deterrent of increased police security in parks and public areas make the tradeoff worthwhile for consumers. In a similar vein, it is growing more difficult to find people who find Amazon's recommended purchases to be an invasion of privacy, or Spotify's music recommendations to be irritating. Many of the same people who are concerned about tracking forget the fact that every time they sit in an Uber ride, they are leaving rich traces of data about your offline trajectory with these companies.

Keep in mind that no improvement to our daily lives, whether it is the automobile, the computer, or our basic utilities, has ever been free of risk. In the case of mobile devices, I believe that the benefits from data sharing, as I describe it in this book, outweigh the risks. If consumers were to open up to the give-and-take, and firms were to respect the intimate relationship consumers are willing to forge with them, the network of providers in the mobile economy can serve the self-selected group of consumers who appreciate advertising, as a butler (and as a concierge), not as a stalker.

It is clear that the advancement of smartphone technology has already exerted a great deal of influence on how we organize our lives. In little more than a decade, it has turned the minds of consumers from a vague landscape of guesswork and speculation to a much clearer field of predictable patterns and preferences we can quantify. The resolution of that field is getting better and better, the more we experiment and learn. But we also need to examine one more aspect of the transformative effect of smartphones: what mobile devices have *not* changed in our behaviors.

3 Striking a Balance

With so many changes in the past decade brought forth by the advent of the smartphone, businesses need to recognize some universal truths that still hold. They ought to internalize the fact that despite the advent of different technologies some fundamental behavioral traits have not changed as much as one would have expected. As we look at these behavioral contradictions it is important for businesses to be cognizant about striking the right balance in consumer lives while leveraging technologies.

The amount of data available to marketers is mind-boggling and will continue to grow as smartphone and sensor technology offers more ways to collect data. The interpretation of mobile data, combined with groundbreaking studies into behavioral psychology, has exposed even more about our fundamental human nature. When it comes to our interactions with the smartphone, we share one desire no matter what country we live in or where we come from. We have the same basic desire for *relevant* messages, which can make our lives easier and our decisions simpler.

On the outside it is very easy to for businesses to get discouraged by consumers' seeming resistance or inhibition to advertising. What lies beneath the surface is different though. On the inside consumers do have appreciation for advertising, albeit at varying levels. This comes forth in four behavioral contradictions, which are particularly striking as we grow more aware of the immense and untapped power of the mobile economy:

1. People seek spontaneity, but they are predictable and they value certainty.
2. People find advertising annoying, but they fear missing out.
3. People want choice and freedom, but they easily get overwhelmed.
4. People protect their privacy, but they increasingly use their personal data as their currency.

Let's take a closer look at each contradiction from the point of view of a consumer.

People seek spontaneity, but they are predictable and they value certainty

There is a reason we call commutes and work-day routines "the rat race." Each day hundreds of millions of adults around the world file into cars, trains, or buses and follow the pack through the maze of streets, tracks, and sidewalks to their cubicles and offices. This drudgery and routine is what spurs workers to take vacations, or to do something different on a weekend or a random day off. We want change and spontaneity, something to keep our minds sharp.

Imagine that something disrupts our routines. Perhaps we get an unexpected afternoon off, which means we can go home early and do something truly spontaneous and positive. How many of us actually use that chance to go somewhere new or try something new? Many would use the time to run their usual errands, or head home and sit on the couch in comfort. We don't always know what to do with ourselves when we have an opportunity for spontaneity. We like change because it challenges us and makes us think, but we also find comfort in routine.

The same holds true for shopping. We like our favorite apparel stores, where we don't have to sample every item because we know how well the clothes will fit. We like to eat at certain places, where we know the food is good and the staff is friendly. While we may try something new, we also like the predictability and comfort of our local spots. "New" for us is often a variation on a theme.

This behavioral contradiction has some roots in biases noted in behavioral economics, in particular the *diversification bias* that stems from our views of how the choices we make today affect our future options. Daniel Read and George Lowenstein revealed the tendency of people to select more variety in purchases when consumption will happen in the future, but more similarity in purchases when consumption will happen immediately.[1]

So how do businesses convince their customers to pause during their day and walk into a store unplanned, or make an unplanned purchase when they were looking for another product? The art lies in making a simultaneous appeal to our comfort and our curiosity. This is one area where mobile marketing shows its inherent strengths. If a retailer knows your browsing history, including shopping tasks you haven't completed because of an inventory stockout online or may have forgotten to purchase after putting the item in the electronic shopping cart, it can let you know when it has an item in stock when you are in the vicinity of the store, and offer you an incentive to come in. If a retailer knows that something you own is due for refresh or replacement, it can make a suggestion knowing that you would be receptive to that kind of advertisement. These spontaneous messages, which seem to come out of the blue are really only parts of your routines and patterns that you are executing out of their normal order. By incentivizing a target customer to enter the store spontaneously, such as a discount or a verified confirmation that the item is in stock and nearby, a business can draw that customer in on a whim.

People find advertising annoying, but they fear missing out

John Gray, the author of *Men Are from Mars, Women Are from Venus*, explains that "to offer a man unsolicited advice is to presume that he doesn't know what to do or that he can't do it on his own." Unsolicited advertising assumes the same thing: that people are incapable of making the decisions on our own. Good advertising, like good advice, can save us time and energy in making decisions or even provide tangible benefits like saving money. Unsolicited advertising, like unsolicited advice, risks being perceived as unwanted and triggering a negative reaction such as annoyance. Thus advertising can create a "social" cost. This annoyance cost is the main reason that advertising and entertainment are often jointly supplied to consumers and balancing the two is an age-old conundrum. Advertising is usually bundled with entertainment because the utility from consuming media content compensates the audience for becoming exposed to ads.[2]

Consumers shouldn't be so quick to dismiss advice (or advertisements) from a trusted source, whether the advice was solicited or not. As the Roman philosopher Seneca said, people are not always the best judges of their wants and desires, nor do they always have all the information: "Consult your friend on all things, especially on those which respect yourself. His counsel may then be useful where your own self-love might impair your judgment." When we have established patterns we may not realize, or we are searching for something but don't know how to find it, a trusted advertiser can take the burden of decision from us and provide us with advice. We simply have to be willing to take their advice.

The challenge for marketers is to turn themselves into trusted advisers, not just sellers of products and services. People often equate unsolicited advertising with bad advertising, especially when the source has not earned their trust. Sally Silver, the chief digital officer of Amplifi, sums up this opportunity well:

> I think it's often a misheld belief that people object to advertising on mobile devices. Yes they are personal, as by their very definition they are carried with us. But in amongst the numerous functions a mobile fulfills—from alarm clock to PA—it is also another device and screen on which we consume media. It stands to reason that consumers will interact with advertising that is of interest to them. Combine that with sequential messaging across devices, and you move past advertising to creating a dialogue with individuals, which is incredibly appealing for brands.[3]

Advertisers need to target users with better ads, with less annoyance. If mobile ads enhance a user's experience rather than merely elicit a transaction, they can convert consumers into strong advocates and positively influence a larger audience.[4] To see advertising more as a dialogue vs. anonymous one-way messaging can ignite a virtuous cycle. Marketers can turn data about consumers into real value, and can then earn more data as a reward.

There is no doubt in my mind that the current ad tech ecosystem has made the mobile Web[5] almost unusable by clogging up consumers' online experience with ads they do not want. This pervasive problem is also an opportunity for the ecosystem to improve.

When customers find advertising annoying, one typical response is to install an ad blocker. They don't want pop ups or ads coming through on their devices, so they "hire" a third-party system to shield them from what they see as an annoyance. The downside of ad blockers is that they treat ads with a "one size fits all" approach. The current generation of ad blockers is unable to distinguish between high-quality and low-quality ads and therefore they penalize all advertisers.[6] Not all ads are low in quality or in relevance. High-quality ads—those that target the right person at the right location and the right time, and align with all nine forces for mobile advertising—are blocked just as harshly as those that aren't as well prepared. Both good advice and bad advice get eliminated in the process.

I believe the advent of ad blockers is a good development for marketers in the long run. In the short run, I can understand why advertisers and publishers bemoan it. But I believe that this threat from ad blockers gives brands, publishers, and advertisers (and the rest of the ecosystem) a tremendous opportunity to make mobile and desktop ads less intrusive and more relevant, and to make them provide utilitarian value.

It is still early days for ad blockers. As of August 2016, only 200 million users of desktop computers and 420 million users of smartphones worldwide use ad blockers.[7] If we take a look at the penetration of desktops and mobile phones worldwide, these are very small numbers. This means an overwhelmingly vast majority of users do not use ad blockers. There is an implicit message in there somewhere from consumers to marketers. An August 2015 survey of US consumers by the Rubicon project revealed that 73 percent of consumers are more likely to engage with online ads when the ads are personally relevant.[8] A survey of users in the United States, the United Kingdom, Germany, and France by Millennial Media showed that a similar picture.[9] Some 63 percent of respondents said they found ads on mobile devices in exchange for free content to be fair, while 25 percent found it unfair. The fairness percentage was almost the same across Germany (65 percent), the UK (64 percent), and the US (67 percent); in France it was a bit lower (56 percent). Consumers are telling firms loud and clear that ad blockers would no longer be a threat if brands would get their targeting right.

Hence, I see an upside to mobile marketers from the entry of ad blockers. Djamel Agaoua, senior vice president of Cheetah Mobile, echoed my sentiments well: "Ad blocking isn't the end of mobile commerce, it's the catalyst that will bring us to the next level. Seen in its true light, ad blocking is a sign that mobile is maturing and a clear indication as to the future of next-level native advertising formats."[10]

Granted that for publishers and advertisers things may have to get worse first before they get better. But that is exactly how a lot of progress in mankind has happened. Facebook's move to block ad blockers and instead let users explicitly convey their ad preferences to the site using the "Ad Preferences" tool is a move toward fixing this problem. It is a remarkably visionary move because it gives users more control over the ads they see and draws them in to participate in this value exchange between businesses and consumers.

I should also mention that many users deploy ad blockers not because they find ads annoying but to protect themselves from malware, viruses, and privacy. According to Jules Polonetsky, chief executive of the Future of Privacy Forum, Facebook's policy of blocking ad blockers does not hurt such users, because it lets them continue to use ad blockers to protect themselves from malware without disrupting the display of ads.[11] So it is a win-win situation.

The mobile device can also do some thinking for us, by connecting or noting occasions we may not have realized yet. There are already apps that sync your location on your mobile device to your chosen music repository. The mobile concert tracker Bands In Town notifies you when a band you seem to enjoy announces a concert nearby, and provides a link to book tickets. This is not intrusive, but rather a service to those who enjoy live music. This same concept can be extended to community service projects or organization gatherings, which now are so numerous that finding and filtering them all would take considerable time. Apps can do that work and send you only information on what is most relevant to you. With greater use of this technology, we can receive more advice we will be sure to take. These are the ads that turn a smartphone into a butler or a concierge. Studies show that a majority of smartphone users want the "concierge feeling" that comes about through high-quality, relevant advertising.

People want choice and freedom, but they are easily overwhelmed

Understanding the third contradiction is as simple as determining what to eat for dinner. You don't want the same thing every night, of course, as that gets boring and tasteless. You want the freedom to choose, and viable options to choose from. But when you look at an app that tells you what is nearby in a big city, the choices can be daunting: Do you want Indian, or Greek, or Italian? Burgers, or pad thai, or sushi? Even if you are in a supermarket and don't have a clear picture in your head of what to make that night, you have all the choices in the world, and, chances are, you will spend more time choosing what to get than you ultimately will spend consuming the meal you eventually make.

In his book *The Paradox of Choice*, Barry Schwartz explained the "paradox of choices" stemming from "choice overload," and why more is actually less.[12] He made the case that eliminating choices can greatly reduce the stress, anxiety, and busyness of our lives. Beibei Li, Panos Ipeirotis, and I found this to be true when we analyzed hotel reservation data from travel search engines such as Travelocity and TripAdvisor. The kind of personalization by hotel search engines that leads to more choices can be detrimental to the firm's revenues.[13] These engines have an "active" personalized mode that allows users to customize the ranking algorithm, and a "passive" personalized ranking system mode whereby users cannot interact with the ranking algorithm. The active personalized ranking system leads to a lower probability of purchase and lower revenues than the "passive" system. Itmar Simonson and Sheena Iyengar did some groundbreaking research in this area of the "choice overload effect."[14, 15]

Providing more information during the decision-making process may lead to fewer purchases by consumers because of information overload and decision paralysis. As consumers, we can process only so many choices at once, and we need to gradually narrow down our options in order to make good decisions. This act does not limit our freedom. It eliminates the feeling of being overwhelmed by our options.

No matter how eagerly we want options, we can no longer make an informed decision when we become aware of all our alternatives. This is a big reason why online ratings websites grew quickly in popularity, from restaurants (Yelp) to product purchases (Amazon) to employment postings (Glassdoor). We want to know what we are getting into. Something that can narrow down the playing field from the full market to a select few that others have verified or validated. The same thinking underpins dating services such as Match.com and eHarmony, which offer tools or algorithms that narrow hundreds or even thousands of potential candidates down to a relevant set with a higher probability of success.

People protect their privacy, but they increasingly use their personal data as their currency

It is natural for consumers to feel protective of their personal information as they hear about threats to their privacy. At the same time, the convenience of storing information such as credit cards, passwords, and contact lists with friends and family is so tempting that we keep the information available and accessible despite the potential risks. We log into multiple websites using login IDs and passwords from Facebook or Twitter, allowing the connected device to grab additional contacts, interests, and even shopping behavior that may or may not be relevant to that app. The burden of having to create a new password prompts many of us to sacrifice our data for the sake of convenience. Put simply, there is a disconnect between our understanding of what it means to be privacy-conscious in the mobile economy and the actions we are taking in the real world.

A large number of consumers are willing to surrender data to get a better retail experience. A survey of users of mobile phones in the US and the UK revealed that 40 percent of them would hand over location data in exchange for targeted goods or services. Likewise, 40 percent would give a summary of their shopping habits in exchange for free products and services.[16] According to Nick Jones, executive vice president of innovation and growth at Leo Burnett, more and more Millennials and members of Generation Z are adopting near-field communication, mobile payments, and similar technologies. They are willing to trade data privacy for personalization.[17] About 51 percent of Millennials say they are willing to share information with companies as long as they get something in return, versus 40 percent of those 35 and older.[18]

But in this age of open information, Millennials are also more cognizant of the privacy risks that a breach in data security could create. The same Millennials and Gen Z members who are willing to use their data as a currency are also likely to be technologically savvier then older generations when it comes to adopting privacy-enhancing measures to protect their personal information. However, there is a disconnect between the value consumers place on their data and brands that trade in data. The Privacy in Perspective study, which was done across the United Kingdom and the United States, sums it up well by saying that consumers are demanding a fair exchange for their data and want to negotiate the terms with brands to mutual advantage.[19]

This suggests that a model will emerge in which there will be an increasing premium for data privacy. When AT&T deployed its high-speed fiber Internet service to compete with Google Fiber, in Kansas it had a very interesting pricing model that captures this concept of a privacy premium. The service was priced at $70 a month to match the price of Google Fiber. But if subscribers chose to opt out of AT&T's "Internet Preferences" program, which recorded users' browsing and search history, they would have to shell out an extra $29 a month.[20] Just because consumers are worried about security hacks of their data does not necessarily mean they want to keep all their data to themselves. Kaveh Waddell summed it up well in an article in *The Atlantic*[21]: "If that's the data-barter economy—give up your personal data and get convenient services in return—then we're seeing the start of data as currency: Give up your personal data and you'll be rewarded in actual dollars."

II The Forces Shaping the Mobile Economy

In this part of the book, we will dive deeper into the nine forces that drive consumers' purchase decisions: context, location, time, saliency, crowdedness, weather, trajectory, social dynamics, and technology mix. Each chapter will take you through the research and analyses that show how companies can confidently and strategically influence one or more forces through new emerging technologies like geo-fencing, geo-targeting, and geo-conquesting. You will encounter real-world case studies that bring to life the concepts that underpin mobile advertising. With the added support of robust research studies conducted using data from brands across industries and geographies in North America, Europe, and Asia, you will gain new insights that you can take away and tailor to your own business goals. The studies are based on real consumers' responses in the real world—subways, shopping malls, online in ecommerce platforms, and offline in physical stores—rather than in the lab. The collective insights will lead to food for thought and recommendations on how to translate data into money in ways that few other strategies and tactics can. When firms can correctly harness those nine forces, mobile marketing does indeed result in higher redemption probability, faster redemption behavior, higher transaction value, higher revenues, higher engagement, higher customer satisfaction, and so on. If you are a consumer, each chapter will provide the insights and examples that show how companies will use the emerging mobile technologies to market products and services to you.

No other platform is better suited to capitalize on all nine of these forces than the mobile platform. It combines targeting with immediacy and context in a way that delivers the consumer the right individualized incentive for the right product in the right store in the right place at the right time.

It is also important to understand these forces as a vast and fast growing body of academic research has shown, however, that under certain circumstances, these terrific intuitive ideas can backfire. The counter-intuitive exceptions to these "rules" can throttle the very consumer impulses they are trying to stimulate. Some behaviors that hold true during the week work less well on weekends. An advertising approach that seems obvious for a certain combination of location and time of day may be less successful or even ill-advised if the target customer's social dynamics change or if the firm is unsure of the consumer's shopping occasion.

With focused investments, companies will not only transform the way consumers view and interact with their products and brands, but will also find the right customers more easily, target them better, convert them faster, reward them sooner, and keep them longer. And companies are creating and improving this reality as you read this.[1] But the business side of this equation isn't the only winner.

When companies challenge themselves and put their insights and data to creative uses, *consumers* can find more relevant products, work with the providers who best match their needs, manage their short-term and long-term "to do" lists, plan ahead, and make better and more informed decisions. And they can do all of this faster, more efficiently, and on a much larger scale than they could have ever imagined even 5 years ago.

This is generating an openness to advertising, especially on something as personal and close to us as our hand-held devices. More and more consumers, especially those under the age of 40, are warming up to the notion that there is a give-and-take relationship between them and the businesses that hope to serve them.[2] Sure enough, there are some differences globally. Consumers in Brazil or China are more likely than those in the United States or the United Kingdom to click on a mobile ad if it is relevant.[3] But even in the United States, the behavior is changing rapidly as consumers become more amenable to interacting with mobile ads. The studies and experiments in this book are testimony to the fact that consumers around the world are willing to take small but steady steps forward in developing trust with marketers, as long as the marketers are willing to move toward the balance mentioned above and provide them real value in return. Give-and-take is what defines these digital relationships and makes them work well for both parties.

A mobile device is an excellent medium for marketing that should maximize the benefits to consumers and minimize the intrusiveness. Consumers have made it clear that if advertisers engage them appropriately on mobile devices it can have a huge impact.[4] Businesses can transform our smartphones to act as our personal concierges—our butlers—and not as stalkers.

Think of some of the apps that already enable your smartphone to serve as a concierge. You start your car, and it proactively tells you the traffic conditions and how long it will take you to get home, because odds are that is where you are headed. If you are tired after a run outside, an app can direct you to a place to buy water or a protein bar. Maybe they remind you when you get near a store that you have gifts cards you haven't redeemed. Perhaps they not only remind you of your flight but also to check the guide to in-flight entertainment so you can plan what you could watch: you pass a subway station and it proactively updates you on delays and tells you when the next train will arrive. These examples are the tip of a large iceberg of what is already a reality.[5]

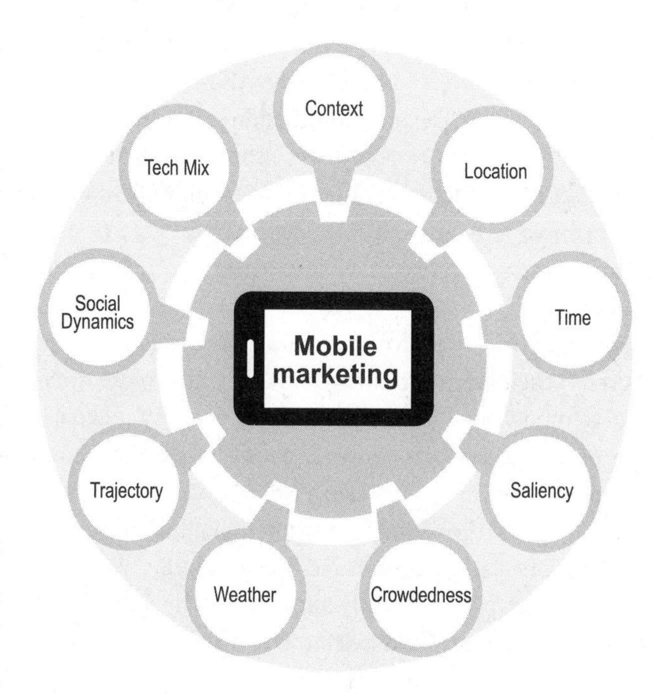

The nine forces that drive consumers' purchase decisions.

It may be easy to draw conclusions about these forces in isolation using some intuition. Classical economics tells us that a company sells more products if they offer a discount. Our intuition tells us that a consumer is more likely to visit a store nearby than one farther away, or choose a purchase option that appears closer to the top of a search or a shopping app. Nicer weather brings out more shoppers, whether they are on foot or in their cars.

The power of mobile advertising lies in the *combination* of these forces. When might a customer ignore an offer from a store that is close by and prefer an offer from a store that is further away? When might a customer ignore a mobile ad that would otherwise be of interest? When sending a mobile offer to a customer, how much time should you give them to redeem it? What kind of targeting will let a business poach its rivals' consumers? Would sending mobile offers to customers in a crowded context increase or decrease their purchase propensities? Does the real-time weather influence effectiveness of mobile marketing? What types of mobile marketing are most effective for driving sales with high-income consumers? Is mobile marketing more effective for single people, or for couples? For larger groups, or smaller ones? Are Web advertising and mobile advertising complements or substitutes? These are just few of the questions I address in this book.

Before we immerse ourselves in the nine forces, we should pause to think about an important aspect of the mobile economy. We are all well aware of the privacy risks that people face as a result of sharing their data or as a result of allowing third parties to collect data from them via their smartphones or other means.

When data analysts look into our lives and mine the data, they may strike a raw nerve, even if their intentions are good. One of the more notable cases is a predictive model that Target developed to identify pregnant women. By tracking purchases across a range of about two dozen items, Target felt confident that it knew not only when a woman was pregnant but how far along she was. It would then send that woman mailings with coupons for baby products. One father confronted Target angrily when his teenage daughter received such coupons. Only later—indirectly from Target and then directly from his daughter—did he find out that she was indeed pregnant.[6]

But for every such sensitive case, one can cite several examples when people are pleased to receive information in exchange for having shared their contextual data. This is an important and huge, untapped form of value creation, accelerated by the proliferation of smartphones and our inability to turn them off or put them down. Companies have the capability and the incentive to re-invest the data they collect into a lucrative marketing strategy, which makes their customers' lives simpler, easier, and less stressful. It's all about context.

But it is profoundly important for businesses to strike a balance. Rob Hammond, Senior Director of Mobile Engagement at Syniverse, summed it up well:

> We need to understand that it's our relationship with the customer that governs what we can do. I like to call it the "creepy/cool" factor. Much the same way that a person can pull away when a casual acquaintance shares way too much information, the customer will withdraw if a brand goes too far past the customer's comfort level. Conversely, when a friend shares something only the two of you will understand, it's a rewarding experience, and you lean in to hear more.[7]

One might be tempted to think that the mobile ecosystem is ripe for abuse, so I include a well-phrased warning from one observer: "Overly aggressive advertising is not the intention with this scenario, but rather kindling better business relationships by providing easier access to what people want."[8]

What is important to make this ecosystem work is for businesses to recognize that mutual respect is a critical element of any relationship. This means that businesses need to understand when and how often to communicate with their consumers. As one observer said, "Whether it's two people or a person and a brand interacting, it's a relationship. Relationships require time to grow and must be cultivated. In other words, just because we technically can communicate something in a mobile message, doesn't mean we should. It's all about context."[9]

That sounds a lot like a concierge—a butler, not a stalker.

4 Context: What's Going On?

Imagine you have to feed 5,000 people for a week and you have one chance to get all the supplies you need. But you can't just offer them anything. These people expect a first-class experience and lots of variety, so you have little margin for error. It would help immensely if you knew something about these 5,000 people.

That is the challenge the kitchen teams face on cruise ships. They look at trends and patterns and need to adjust menus on the basis of route, season, and passenger profiles. They even install ceiling cameras to track the flow of guests and plan for peak dining periods.[1] The breakdowns and analyses go even further. For instance, the nationality of the guest matters. More Germans or more Koreans onboard means you will need more pork. More Americans means you will need more chicken and beef. More women means you will have to serve more Caesar salads.[2] Americans prefer chewy cookies, Europeans crunchy ones.[3]

Then you have to plan for beverages, especially alcoholic ones. It helps to know that Europeans drink different wines than Americans, but the pattern analysis goes into even finer detail. Some cruise lines look at over the last 2 months and at year-on-year comparisons, and make adjustments for the type of cruise ship (luxury vs. more accessible) and for the season.[4] The age of the guests matters too, and the cruise ships must plan for special events that may take place during the cruise (sports championships, for example) and thus may change the guests' demand profiles.[5]

All this coordinating is a daunting task, aimed at providing the guests the kind of positive and memorable experience that will encourage them to spread the word and come back again. The cruise planners are endeavoring to understand and adapt to what I refer to as "context." It is the sum of all factors, circumstances, and associated behaviors that guide decision making, whether we are customers or the businesses that serve them.

What makes it possible for the cruise planners to excel at this? They have the benefit of access not only to historical customer data but also to data on who their future customers are (for the imminent trips). While many businesses may have lots of historical customer-level data (for example, from loyalty cards), they do not easily have information on their imminent customers. But this is exactly where the mobile channel can be an exceptionally useful strategic and tactical tool. By leveraging various mobile technologies, businesses can incentivize customers to walk into their store at any given point in time. But they can do more. By combining their historical data with real-time signals from mobile devices, businesses can influence *who* they want walking into their stores!

From the business perspective, the territory within a certain radius around your physical location is like your own cruise ship. The better you know who is in the area, and who is "on board" for your brand or your category, the more you can do to attract them, make them feel welcome, and hopefully do business with them and earn their loyalty. But how do you know who these customers are?

In the pre-Internet era, answering that question, even crudely with some guesswork, would be the best a business owner could do. They could look at demographics and make some superficial judgments based on age, gender, general appearance, and how much the customer spent. As we progress through the 21st century, when nearly every potential customer roams the streets with a smartphone, answering that "who" question is no longer the best we can do. It is barely even a good start.

Our Different Avatars: Digging into Our Modes and Moods

In the mobile economy, knowing "who" your customer starts with identifying each customer's multiple personas. These depend on day of the week and time of the day. Consumers' personas or modes vary depending on the context. When I'm taking my daughter out for self-defense martial arts classes on Sundays, I'm in the "parent mode." When I am out on a business meeting with clients in my consulting practice on a Wednesday evening, I'm in the "work mode." When I am at the airport lounge waiting with my family for the flight to take off, I'm in the "family mode." When I am in the parent mode, an ideal offer for me is a discount on a cappuccino, given that I need to spend the next hour watching my daughter take her lessons. In contrast, when I am in the work mode, an ideal offer would come from a pub promoting discounted drinks.

Today, the "Who" your customer is no longer just about understanding the different personas of your customer; it is multi-dimensional. Today, using context to your advantage means knowing, for every customer, the answers to three questions:

- Why is the customer there?
- What does the customer want now?
- How is the customer feeling now?

These questions sound deceptively simple. But as we take a closer look at them one by one in this chapter and in the remaining chapters of part II, you will see how multi-faceted they are and how powerful their answers can be, especially when we look at how they interact together and with the other eight forces. This makes "context" a foundational force. The remaining eight forces build on it to create economic and social value for consumers like never before.

Think of this simplified example of context. A person walks every weekday morning to a chain coffee shop for the "usual" and then takes a 20-minute subway ride to work. If the weather is nice, the commute is a walk or a run, with coffee coming later. Lunch comes from any of a few takeout places within a couple of blocks from the office, except on some Wednesdays when a few friends get together for a longer, sit-down lunch. When there is a lot of stress at work, though, lunch is always the same: order in, shut out the world, and plan on working late. Once or twice a week, that same subway line takes our office worker to dinner, a show, or a sports event with friends.

The person in that example could be a 50-ish manager at a standing desk in a corner office or a recent college graduate hunched in a cubicle. That manager could be a divorced mother of two college-age kids or a man of the same age who has never been married. They do have some activities in common, though, such as running. Under Armour is one example of a company that has picked up on these differences among people who undertake similar activities at different stages in their lives.[6] The company uses data collected from wearable devices to keep its product line relevant as customers' lifestyles or fitness levels change. In 2013 the company purchased MapMyFitness, giving it access to data across a platform with 20 million registered users at the time of purchase.[7]

The truth is that we all live our lives in different moods and different modes. What we eat and drink, how impulsive we are, how active we are, and what we search for and expect to find all depend on our mood. We celebrate, we eat comfort food, we indulge ourselves, or we abstain, all depending on how we feel. At the same time our behaviors and decisions depend on whether we are in "work" mode, "parent" mode, "family" mode, or any number of other situations we often find ourselves in.

Consumers' long history of browsing and buying online, followed by the last decade or so of having smartphones infiltrate so many aspects of our lives, has generated a trail of data that reveals patterns, preferences, and motivations that until now remained under the surface and were addressed partially only through educated guesswork. This set of data grows richer and deeper every time we switch on our smartphones. The more consumers are willing to give businesses access to such data, the more efficient businesses can make their lives—even if those same businesses also profit from the exchange.

We are all familiar with the smartphone as a means of two-way communication. We text with friends, send and receive emails, interact with friends on social media, exchange information with colleagues, and conduct our personal business with many different vendors. Occasionally we may even make an old-fashioned phone call.

Another form of two-way communication is the exchange of information about consumers with third parties, who can respond with ways to make your life easier. I realize that this requires a new comfort level for many people, but over time I believe that the benefits will outweigh the risks and offset the fears once consumers have realized just how great the benefits are. Consumers can save time and money, and gain satisfaction and peace of mind, if the whole range of people and businesses who can serve them—from stores and restaurants, to professionals and tradespeople, to entertainment and travel providers—know more about them and their context at any given moment.

In the remainder of part II, we will examine the other eight critical forces behind our purchase decisions that make mobile advertising so powerful. These forces—location, time, saliency, crowdedness, weather, trajectory, social dynamics, and tech mix are derived from how we answer the three basic contextual questions above. But before we get to the remainder of part II, we will take an initial look at these contextual questions and the role our mobile data can play in answering them.

Why Is the Customer There?

This is a combination of contextual elements, including psychological ones. Consumers' motivations affect how they make purchase decisions and how they respond to incentives. These distinctions can be as basic as planned purchases vs. impulse purchases, whether they are gathering information from a website, or a floor sales representative, or whether they are making last-minute purchases. More subtle distinctions include aspects such as urgency, who recommended making the purchase (friend, family or spouse), and the consequences our decision may have.

Think about it from a consumer's perspective. Buying six-packs of beer, buying backpacks, or buying new outfits for your wardrobe never truly involve identical transactions, even if the item itself is essentially the same. When you want the beer, is it for yourself to have on hand, or perhaps for a game night you are hosting? Do you have friends visiting from out of town, or will you join some friends tonight for a hastily arranged party to watch a playoff game? You may need a backpack for a one-week hiking trip this summer, but would consider buying it sooner rather than later if a sporting goods store or a specialty store offered you the right incentives. You may need a new outfit because you are starting a job in another city four weeks from now, or because you are meeting your life partner's family for the first time.

All these factors help define the occasion. They can affect everything from the brand a consumer buys, where they shop, their willingness to pay, and even their desire to "get it right" and make a good impression.

Imagine the possibilities if the smartphone served as a two-way channel with businesses. When a customer needs that new outfit, how different would her life be if she could walk down Fifth Avenue in New York City and exchange information—about size, style, colors, occasion, and perhaps even budget—with stores before she ever walked into anyone of them? If she chooses between Saks and Bloomingdale's, either because of their advance service or, say, because of the coupon they sent her, wouldn't it be wonderful to walk straight into the store, head to the right department, and have three ensembles already waiting for her in a changing room, ready to try on? Sharing personal information in this manner turns your smartphone from a tool you need to engage with into an active pocket butler that puts others to work on your behalf. Of course, I must add that shopping is also a fun activity for some people this efficiency is not a relevant factor for people who love to browse!

Businesses in turn have many possibilities to improve their services and influence the most valuable customers when they have access to the kind of contextual data about who visits a different location at different times. This can be especially relevant to event-based marketing.[8] Imagine the various concession stands during an event in Madison Square Garden. Because the management of the venue can pinpoint the interests of different demographic segments who visit Madison Square Garden depending on the events held there, it can customize the kind of merchandise and food it sells depending on whether it is hosting a Knicks or Rangers game or a Justin Bieber concert. Because the context of who visits Madison Square Garden during those events differs radically, the management can benefit from creating content that demonstrates that they get their audience.

What Does The Customer Want Now?

The needs consumers are trying to fill also contribute to context. They have both psychological and practical aspects. Whether it is morning coffee, ice cream on a summer day, or a drink after a tough week, most consumers smile when they hear the question "Your usual?" That brings comfort, satisfaction, and faster service along with the product itself because the staff saw you walk in the door.

What many consumers fail to realize is that they have many more "usuals" than they think. Their shopping for everything from groceries to furniture to entertainment and travel usually follows certain quantifiable and predictable patterns. Because search costs and transportation costs are often high in the physical world, where and when consumers move around can indicate their preferences and habits. It also exposes what their other "usuals" are. The more you as the marketer know about the range of your customers' "usuals," the better you can serve them, even if you are meeting them online or offline for the very first time.

Let's go back to beer and backpacks. When a consumer is buying a backpack, they want more than the object itself. They may also want service, such as informed recommendations, a smooth and friendly shopping experience, and easy delivery and return policies. They may also want other ancillary things for the trip besides the backpack, such as water bottles, extra clothes, poles, or glasses.

Each consumer has some expectations about the quality and type of service they receive. The common offline situation in a store is sometimes an arduous cat-and-mouse game of twenty questions between the customer and the salesperson. You may be the kind of person who searches first and asks later, fearing that the salesperson only wants to upsell you. You are already frustrated that you couldn't find what you wanted on your own. Or maybe you are someone who seeks out a salesperson right away. You fill them in on the details of your trip (destination, activities, duration, anticipated weather, size of the group) and they need to respond spontaneously to your questions.

Online is a different story, as we will explore in the chapter on saliency. A consumer will conduct a search for her backpack on a website or through a mobile app (or a combination of both) and need to scroll through dozens if not hundreds of results on a small, two-dimensional screen.

In both cases, some form of advance communication about what she wants (the product) and how she wants it (the service) can signal stores in the area and encourage them to compete for her business and "right sell" her. Imagine if a store knew in advance not only about the current trip she is planning, but also the previous hikes she went on, or even knew that she has a big adventure on her bucket list such as hiking through Nepal or climbing up to Machu Picchu.

New considerations: Using trajectory to extract the "usuals"

The easiest way to extract information on these "usuals" is to start with the past. Consumers have a history of purchases that reflect and reveal their habits and tendencies. Over time, they leave a record of where they have been, how long they have stayed there, what they purchased, how much they paid, and when they purchased it (time of day, day of week, month or season). Such a record exists in the online and offline worlds. I refer to these past movements in the offline world as a consumer's trajectory.

A user's individual mobile trajectory is to location-based advertising what a video is to a photo snapshot. If a picture is worth a thousand words, what is a video worth? Snapshots also take contextual information into account, but they cannot reflect or reveal patterns or tendencies in the way a "video" or "trajectory" can. A mobile trajectory can include what locations a consumer has visited, when, for how long, and what the overall context was. Measuring this trajectory is important for understanding a consumer's inherent preferences and creating two benefits from it: a superior customer experience and a better marketing strategy. We will look at "trajectory" in depth in chapter 10, as it is an important force in its own right.

The future may be the most fascinating and perhaps the most lucrative part of "context" is its ability to project the future for businesses. When we project someone's trajectory into the future, or match someone to an expected trajectory on the basis of who the person is and what he or she is currently doing, we can anticipate what someone will likely do today, tomorrow, a month, or even a year from now. Recall one of the behavioral contradictions from chapter 1: Consumers believe they are very spontaneous, but they are in fact the sum of their "usuals" or a variation on those themes. That sounds almost too good to be true, so a team of researchers in Seattle decided to test out that idea. They analyzed more than 32,000 days' worth of GPS data across 703 diverse subjects and demonstrated that it is indeed possible for a firm to predict where you will be months and even years from now![9] We are far more predictable in our behavior than we think, and those of us who are constantly tapping on a phone are creating an excellent source of data on our traveling patterns.

How Is The Customer Feeling Now?

This covers a customer's state of mind and their social dynamics. On paper, who the customer is won't change between today and tomorrow. A customer probably won't update any basic profile information on a website or change settings on his smartphone. But tomorrow he might not feel well, or could be outright sick. If he is feeling good, he may go out with friends, with his partner, or with his children. These are "modes" such as were spoken of at the beginning of this chapter. As one may expect, consumers change their behavior depending on how they feel and who they are with.

People as couples respond differently than they do when they are alone or in large crowds. Surrounding themselves with people changes how people feel and how they act.

As I say over and over again in the book, it is unwise to make a blanket application of any of these statements. Context involves many variables, and a business must take all of them into account to get a complete picture of a potential customer's receptiveness and responsiveness to mobile advertising. One example illustrates this point. Using one of the largest shopping malls as a context, my colleagues and I showed that, in contrast to people with average incomes, high-income customers are more likely to respond to mobile ads when shopping alone than when shopping with others in social groups.[10]

New considerations: Micro-moments and in-the-moment marketing

This knowledge of "how" has a bearing on the present, when a business and a customer face today's moments of truth. Today a consumer probably had several moments of truth, large and small. Google refers to these as "micro-moments." Every time you search for something or pay for something, it gets a time stamp. A lot of these micro-moment searches are "local," and purchases are highly likely. Marketers need to be there at the right time when the consumers need help, rather than only spending in brand awareness that often lacks the focus on immediacy. Mobile will enable brands to offer this kind of in-the-moment support. In anticipation of consumers' searches, services such as Apple Spotlight and Facebook have customized, time-specific recommendations ready even before you begin a search.[11] Based on the time of day, Apple's Spotlight has suggestions for popular searches for cafés, gas stations, delicatessens, etc. With a single touch, it takes potential customers to its Maps interface where they can see the exact locations of these stores. Apple and Facebook are just two examples of several firms who are painting such contextual marketing masterpieces.

In a recent study on the effectiveness of contextual mobile marketing, Panagiotis Adamopoulos, Vilma Todri, Alexander Tuzhilin, and I analyzed consumers' responses to various types of recommendations in a mobile app.[12] These recommendations were about local venues such as restaurants and bars. We had data from 12,119 venues across the ten most popular cities in the United States and the corresponding visits of several million active users of one of the largest US-based mobile urban guide apps. We found that consumers were significantly more influenced by the "in-the-moment" recommendations about currently trending venues than by experts' recommendations or recommendations based on historical data. A 10 percent increase in the frequency of "in-the-moment" recommendation raises demand by 7.1 percent, clearly demonstrating the power of contextual mobile marketing.

Several firms across the world are harnessing consumers' contextual data to deliver relevant messages and unlock the potential of the mobile economy. An electronic discussion with Ankur Warikoo, the CEO and co-founder of Nearbuy (formerly Groupon India), led me to understand that Nearbuy is an example of one such firm. Nearbuy tries to infer the why, what, and how of each consumer by combining information on the user's location with the time of the day and the behavioral history of the user. They know what the user has browsed, liked, bought, abandoned, and shared in the past. They also know the venues the user frequents (and also where the user's office and home are). By cross-referencing consumers' behavioral patterns with venue profiles helps them create contextual data on where users and when. So when a user enters a specific tagged location in Bangalore, Delhi, or Mumbai, their app triggers a notification to the user with a marketing offer. The notification is customized based on the time of the day. If it is morning, they may prompt users with breakfast options or shopping options. If it is afternoon, the offers may consist of lunch, spa, or movie options. If it is evening (say, a Friday evening), they may prompt the user with options from pubs, clubs, and restaurants. By combining the location, the time, and the user's behavioral history, they are able to achieve a very impressive 30 percent click-through rate (as compared to 4 percent for a generic notification) and a 12 percent transaction conversion (as against a 7 percent a generic notification)

Food for Thought

The recipe for success in mobile marketing is simple: consumers contribute real-time contextual data, and businesses use such data to generate and deliver better value. So why does this symbiotic relationship between consumers and marketers still seem a bit fragile? A better way to phrase that is to ask whether the smartphone and its kin are opening the door for stalkers or paving the way for butlers and concierges to solve our problems and make our lives easier.

Consumers are a bit uneasy, because their data are the fuel that powers the mobile economy. Without it, all the techniques and approaches I describe in this book become little more than "What if?" and "What could have been?" theories. Businesses are also treading lightly, as they move along a spectrum of acceptance that goes from "Great idea!" to "Wow, this really works!" to "We have to do this!"

As we will see in detail in the remainder of part II, businesses have a disincentive to bombard their potential customers with advertisements, wanted or unwanted. They have to find the right balance and be selective about whom they target. The objectives of both parties are a positive experience for the customer and a profitable transaction for the businesses. Poor or incomplete information flow is a barrier to that kind of experience, and the smartphone is a floodgate. Open it and you reduce or remove that barrier.

The Chicago blues musician Anthony Moser penned a song called "Google Is Listening" that lists a litany of activities he thinks Google knows all about. An aptly titled article in the *Boston Globe*, "Why our location histories are a glimpse into the future," highlighted this tension between what companies such as Facebook and Google collect and what they do with it—the article's author agreed to surrender "his coordinates and some battery life" and was surprised to see that Facebook not only could tell him what friends were nearby but also could trace almost all of his movements while he had his smartphone on. He saw the benefits, the risks, and the resulting tradeoffs.[13]

Thus, for consumers, opting in might mean letting go of their data. But it does not mean having to compromise on their data privacy. It means opting into the whole idea that what one finds on one's smartphone can make one's life much easier. This will continue to improve in the coming years. As consumers enter into a give-and-take relationship with firms, they will give firms their data; in turn, firm will make them offers relevant to their context. It pays to be demanding about what they want brands and retailers to deliver.

Takeaways for Firms

Consumers are increasingly willing to uphold their end of the bargain and share their information. The mission for businesses is to treasure this trust as an asset and reward their consumers with relevant and non-redundant messages. Relationships require time to grow and must be cultivated. Businesses today are in a position to know things their predecessors had to infer. The best way for businesses to capitalize on this large and growing amount of knowledge is to keep the following things in mind:

- Identifying and leveraging the "usuals" (habits, preferences, behaviors) of its customers is the first step toward devising a contextual mobile marketing strategy and unlocking the potential of the mobile economy.
- Data from mobile devices can help a business recognize the distinctions between the historical (why), personal (what), and psychological (how) context for a given consumer. These three elements form the core of a contextual mobile marketing strategy.[14]
- Just because a business has the technological capability to communicate with a potential customer at any given point in time, doesn't mean it should. It should wait for the right "micro-moment."[15] It should focus on creating less frequent but more relevant messages. This patience can make its messages truly contextual so that consumers perceive the device as a butler and a concierge, not a stalker.

5 Location: Why Geography Matters

Think about how many places you visit or frequent. Most people have a home, a workplace, a home town, a small set of establishments they visit often, routes they routinely follow, and destinations they travel to, both near and far.

Even when we narrow "location" to one specific place, our perception of that locale can vary greatly. These differences will manifest themselves in the way we describe and define that location. Think back to the individual in the previous chapter who—according to GPS—is standing on the corner of Fifth Avenue and 49th Street in New York. Let's say her name is Linda, and her phone beeps. She looks down and sees one of the simplest questions she could receive as a text: "Where are you?"

Definition of Location

Now think of all the ways she could answer that question:

• "I'm on the southeast corner of Fifth Avenue and 49th Street." This is her most precise geographical location in a literal sense. This is the dot she would see if she would call up a map app and search for something.
• "I'm in midtown." This is synonymous with being in the heart of midtown Manhattan. But it provides no indication of where she might be headed, what brought her to this location, or how long she may stay.
• "I'm in Manhattan." This answer is helpful, but is not very precise. Geographically speaking Manhattan is a large area (roughly 22.8 square miles), and many people will offer a major city (e.g., New York, San Francisco, or Los Angeles) as an answer when they mean the metropolitan area, not a place within the city limits.

• "I'm a block from the 51st Street station." She is probably planning to get on the number 6 subway train and leave that immediate area.
• "I'm standing next to Saks Fifth Avenue." She uses the closest retailer as a landmark to describe where she is standing.
• "I'm across the street from Rockefeller Center." She uses another building as a landmark, but not the one in her immediate vicinity
• "I'm a few blocks from Radio City Music Hall." The description defines her location in terms of both a landmark (again, a prominent building), combined with a distance.
• "On my way home." This interprets the question as "Which stage of your routine journey are you in?" rather than a request for a precise location
• "I'm 20 minutes from the restaurant." This description combines the time element with a very specific, mutually known location not in the immediate area.
• "I'm not sure." This answer implies a lack of familiarity with the area, even with its prominent landmarks.

The list could go on and on, but you should notice that the idea of location is far more than what the GPS coordinates show. We can define it and describe it using landmarks, time, distance, direction, and motivation, or a combination of any of these factors.

It is the richness and variation in these descriptions that make location-based marketing such a compelling opportunity for businesses. In fact, how a consumer chooses to answer the question "Where are you?" can offer rather precise clues about their receptiveness to advertisements and what kinds of advertisements (push or pull) work best. These descriptions can be very advantageous to customers, allowing them to attract the relevant guidance to accomplish their given tasks more efficiently or more affordably.

It is the richness and variation in these descriptions that make location-based marketing such a compelling opportunity for businesses. In fact, how consumers choose to answer the question "Where are you?" can offer rather precise clues about their receptiveness to advertisements, what kinds of advertisements (push or pull) work best, and what they offer. At the same time, these descriptions can be very advantageous to customers, by allowing them to attract the relevant guidance to accomplish their given tasks more efficiently or more affordably.

For many decades, location-based marketing, at best, meant being able to target users at the level of ZIP codes. Every consumer in the same ZIP code received the same offer, presumably because they had homogeneous preferences stemming from similar socioeconomic and demographic backgrounds. When Nike released a new NBA-player-endorsed product, such as the latest LeBrons or Air Jordans, it would target radio and television ads at certain ZIP codes.

In today's mobile economy, brands can target a dramatically finer level of granularity. Not only can firms know in real time which store someone is visiting; they can also know which aisle in the store someone is standing in, and in some cases estimate shelf location of the product someone is staring at. This can be done when sensors installed in LED light bulbs can act as navigation devices. Each bulb is equipped with visible light communication (VLC), enabling it to beam out a code that's imperceptible to the human eye. A forward-facing camera and a smartphone app read the VLC. Phillips has been pioneering this in supermarkets in Düsseldorf and in in France.[1] Purely from the perspective of location, it doesn't get any more granular than that.

I am not implying that any of the retailers shown on the map in figure 5.1 can eavesdrop on the private text messages of every passerby. But it is possible for some retailers to make inferences about Linda's intent and also her needs at that time from a combination of her location history and her recent history of browsing with their mobile apps. Her location and her own description and perception of it help determine whether she is a more or less an appropriate target for a targeted mobile offer. Let's assume that she is in New York City to purchase an outfit to wear to her first day on a new job. How she thinks about her location and how she describes it, in response to that text, can vary greatly, even if her feet remain planted in exactly the same place.

Geo-Awareness: A Fundamental Part of Any Advertising Strategy

The answers to "Where are you?" underscore the complexity of planning and executing an advertising campaign that includes location-based advertising.

Consumers' location histories are in fact very predictive of their product preferences. This means their locations will invariably influence their responses to marketing offers. A recent study by Peter Zubcsek, Zolt Katona, and Miklos Sarvary validates this insight. Using data from a mobile

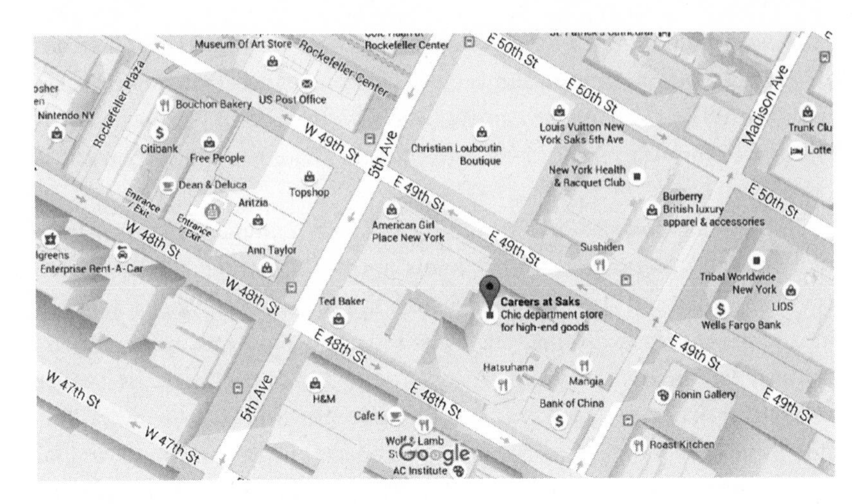

Figure 5.1
A map of the area around Fifth Avenue and 50th Street in New York. (Google Maps)

operator in the Pacific region, Zubcsek et al. showed a positive relationship between where consumers were located and how they responded to mobile promotions, even after controlling for demographic and psychographic differences.[2] Consumers who frequent the same locations have similar preferences, even though they may not know each other. In chapter 8, where I talk about "crowdedness," another example of this behavioral similarity will be presented.

The total value of global real-time location-based advertising (LBA) is expected to grow to $14.8 billion in 2018 from $1.66 billion in 2013, which is a compound annual growth rate (CAGR) of 54 percent. LBA would then represent 39 percent of all mobile advertising and marketing.[3] Location-specific advertising (e.g., coupons) can be pushed to consumers by making use of their location in the form of a "mobile push." Push advertising means that the advertisement seeks the customer, with the available technology "pushing" the coupon, advertisement, or other message onto the customer's device … ideally at the right place, at the right time. Push-based mobile ads can be delivered via text messaging or via a notification within a mobile app. Since 98 percent of text messages are read within 90 seconds of delivery,[4] text messaging can be a very fast and effective tool for a brand to communicate a real-time offer to its customers. Bohemian Guitars, IHOP, E-Trade, American Airlines, and Bloomberg have achieved dramatic results from text-messaging campaigns.[5] Although text messaging is extremely effective for simple messages, rich media messaging (RMM) is increasingly used to send video-based and image-based offers. Express, Avenue, and other brands have achieved extraordinary success in redemption rates and returns on investment with RMM mobile offers.[6]

Alternatively, location-based advertising can be rendered in the form of a "mobile pull" within an app where users can purposefully browse through the available coupon promotions. Pull advertising in the mobile context means that the customer seeks the advertisement, either directly or indirectly. The consumer uses an app (either an aggregator of merchant offers, like Retailmenot, or an individual merchant app, like Target) to find advertising or promotions for a particular product or service category. They pull the information to themselves. The fact that the consumer wants to have a list of offers to review implies that it is highly likely that they have a desire to make a decision and a purchase.

A number of companies have demonstrated strong results with location-based targeting.[7] A campaign by Dr. Pepper and its marketing partners drove over 200,000 store visits and put Dr. Pepper products into 25,000 new households. The Monterey Bay Aquarium worked with Apple to add indoor positioning to its existing app. They mapped 200 galleries so visitors can now use Wi-Fi to access them via a branded app. This lets visitors access a guide on the precise locations of more than 35,000 animals and plants that are on display exhibits. Memorial Healthcare System, a provider in South Florida, wanted to identify new patients to get more leads to promote its new cardiac risk assessment, so it worked with the music streaming service Pandora to target cross-platform ads on Web, mobile phones, and tablets to on-the-go people, especially those over the age of 40. It used geographic and demographic targeting to reach out to the right potential audience. The campaign resulted in a 500 percent increase in mobile-phone-based traffic. In Norway, Coca-Cola worked with the proximity marketing platform Unacast and with a media partner to target moviegoers in specific locations with an indoor campaign. Users who had a specific app from *Verdens Gang* (a major Norwegian newspaper) were asked before the movie began if they wanted a free Coke. Around 60 percent of customers clicked through, and 50 percent redeemed the offer—extremely high rates, given that the average click-through rate on its historic mobile ad campaigns was 0.18 percent.

LBA is growing rapidly as a share of mobile advertising, and for good reason. Knowing where a potential customer is—or how far away the customer is—enables businesses to consciously select the right tactic and execute it the right way. The overarching term for this knowledge is geo-awareness. The breadth and depth of available data continues to grow. The data available through the crystal screen won't just create these possibilities. The data will enable marketers to execute and calibrate them. Once a business has developed a level of geo-awareness, it can pursue tactical options, which fall into three broad categories: geo-targeting, geo-fencing, and geo-conquesting.

Tactics for developing a geo-awareness strategy

Geo-targeting is the most basic form of location-based targeting. Geo-targeting means that a business knows—or assumes—the location of a website visitor from the IP address, the ZIP code, or other information. This knowledge allows the business to provide customized information for that local area, taking that area's own context (current weather, prevailing tastes) into account.

Geo-fencing triggers a signal when a potential customer physically enters a defined geographical area. Think of Banana Republic and Bloomingdales in New York City. One of those stores may define a specific radius around the store (say, half a mile) within which there is a reasonable expectation that someone may alter or complete his journey and come to the store if he received the right incentive. Remember Linda who was standing at the corner of Fifth Avenue and 49th Street in New York? The place where Linda is standing is less than a fourth of a mile from either store, so if they were deploying a geo-fencing solution, Linda would be in the vicinity.

Geo-conquesting takes the idea behind targeting and fencing one aggressive step further. It means that a business will send a potential customer a coupon when that customer enters or approaches a rival store, in an effort to lure them away and win that customer.

Geo-fencing can also be used advertisers to figure out when consumers are within the view of an offline ad (such as a billboard ad), and then attribute offline advertisement impressions to online sales. Two examples cited in a recent report by Skyhook, a mobile location services company based in Boston, illustrate this point.[8] Imagine a large retailer like Macy's purchasing a billboard ad in an airport terminal or on a highway, either of which could be a pre-recorded geo-fenced location. When a Macy's app user passes the billboard, the mobile app will record that he entered the geo-fence around the billboard and connect that information to subsequent online purchases made by that user. Alternatively, it could also track when the user enters the Macy's store in New York City, and on the basis of that information the billboard advertiser would receive credit for driving a real-world conversion.

One of the earliest examples of a business pursuing a geo-aware strategy was Staples. In 2012, the *Wall Street Journal* reported that the website of the office-supplies retailer Staples was offering different prices for the same product on the same day to different customers after estimating their locations.[9] More specifically, if a rival store such as Office Depot or OfficeMax was within 20 miles of the customer's location, Staples would offer a discounted price. Further, Staples also took average income by location into play in its pricing strategy. Locations that tended to see the discounted prices had a higher average income than areas that tended to see higher prices. The same article reported an acknowledgment from Office Depot that it uses customers' browsing history and their computers' IP addressed to vary the offers it displays to a visitor to its site.

With the rapid adoption of LBA, it is becoming imperative for businesses to build their expertise in this space. What are some of the factors driving the location-based mobile economy? My colleagues and I have conducted several studies to examine this question in ways that can help firms realize non-trivial benefits from engaging in this practice.

Conventional Wisdom on Distance and Discount

Studies have supported from many different perspectives the idea that distance between the user and the brand is an important driver of advertising responsiveness, not just location itself. Consumers are more willing to act on a promotional offer if the event is nearby[10] and discounts can literally draw customers closer to the store. This applies to the mobile ecosystem as well and is the foundation for geo-targeting and geo-fencing. The notion is intuitive and almost obvious that the further away a customer is, the less likely he or she will be to make a purchase. The idea that one can understand, measure, test, and capitalize on the relationship between distance and discount opens up myriad possibilities for marketers to create the right purchase incentives for potential customers.

One field experiment established a clear functional relationship between distance and the likelihood of redemption, at least for restaurants.[11] The further away customers were from the restaurant upon receiving a coupon, the less likely they were to redeem it. Consumers also rely on location-based apps to coordinate with friends and acquire local information.[12] Combining these results with the widespread use of GPS-enabled technology, one can conclude that a marketing campaign's chances of success improve significantly when it can target individuals close to the potential point of sale. That is exactly what Avi Goldfarb, Sangpil Han, and I set out to understand in a 2013 study. Our study validated that consumers engage with ads for local stores in close proximity to their homes at the time they conduct their mobile-based Internet searches.[13] Specifically, we had looked at user impressions and engagements for a microblogging service. Similar to Twitter, users could post, read, and respond to posts of up to 140 characters, with the option of clicking through to read more text or see other related media if the text exceeded the 140-character limit visible on the screen. The brands ranged from relatively unknown ones to globally familiar brands such as McDonald's and Starbucks. For both types of searches (PC and mobile phone), the likelihood of clicking on a post increased as the physical distance to the respective store decreased. Consumers prefer stores that are closer to them, whether they are sitting at home or in the office, or underway with their smartphone in hand. But the most important takeaway for mobile marketers is that the difference was twice as strong for the mobile search as for the PC one. For every mile the distance between user and store shrank, the likelihood of clicking on the post increased by 12 percent for PC users and by 23 percent for mobile phone users.

Counter-intuitively, this study showed that distance matters even more on a mobile phone than a PC! So if a business were to target a customer, in order to incentivize him to come to its store, it would have to give twice as much of a discount on a mobile channel as on the PC channel.

New considerations: Distance vs. discount tradeoff

Let's focus now on quantifying the relationship between distance and discount and the related tradeoffs. One large-scale study yielded precise and powerful results. Dominik Molitor, Philip Reichart, Martin Spann, and I conducted a study in collaboration with a leading telecom provider in Europe. Our study was based on data from more than 3,544 different stores spread across 374 cities and towns in Germany and was conducted over a period of 14 weeks.

It is not surprising that the further away the stores' locations are from the consumer's location, the lower the probability that a consumer will choose to redeem a location-based coupon in the form of a pull advertisement delivered via an app. But this study showed us how any distance disadvantage can be compensated for with money or discounts! We showed that one additional kilometer distance between the user and the store decreases the probability of choosing a coupon by between 2.0 percent and 4.7 percent. However, one percentage point increase in discount is the same as *decreasing* distance by between 92 and 230 meters between the user and the store. A business can literally use money to shorten the distance between a customer and the store, or put another way, to offset a distance disadvantage.

If Linda has a particular jacket in mind, she can stand on the corner of Fifth Avenue and 50th Street and conduct an online search. If she sees the same jacket at Saks Fifth Avenue (less than a block away) and Bloomingdale's (half a mile, roughly 800 meters, away), a coupon from Bloomingdale's offering a discount of 10 percent or less can offset its distance disadvantage relative to Saks. It will encourage Linda to head there rather than to Saks.

But this is where the perception behind Linda's answer to "Where are you?" matters. If she says that she is "one block from the 51st Street station," it may indicate that she has no intent to stop at Saks Fifth Avenue at all. It may be on her list of stores, but probably not at or near the top.

Can a Discount Always Compensate for Distance?

New considerations: Direction of travel

Now let's look at Linda's situation from another perspective. Suppose that Linda plans to buy an outfit, but that at this moment she is on her way home for the evening. What also matters in our decision making is the typical direction of travel that a consumer normally follows. When a consumer receives an offer that would take her away from her normal commuting direction, she may not appreciate the offer. This holds true even if the store that sent the offer is closer than a similar or identical store in the direction of their commute. In short, if Linda is heading home at the moment, she may find a coupon from Saks Fifth Avenue or Bloomingdale's, which are located in the opposite direction to her commute back to work, to be a distraction rather than a useful form of guidance, even if she plans to buy an outfit in the immediate future.

In a recent study, Eric Kwon, Dongwon Lee, Wonseok Oh, and I worked with the largest provider of mobile services and the public transit system authorities in Seoul to show that the average of our commuting patterns is a much better predictor of our responsiveness to mobile marketing offers than the variance in our commuting patterns (as measured by days when we are not undertaking our regular commutes). When on the go on a routine day, we prefer the comfort of our own "rules" rather than exceptions or diversions. When we deviate from our usual traveling patterns, we are less likely to be responsive to mobile marketing than when we follow our normal commuting patterns.

This means that a business situated in the usual direction of travel that its target customer is going to undertake can incentivize the customer to come to its premises with a discount smaller than the discount a business in the opposite direction of usual travel would have to offer. Knowing your customers' regular traveling patterns can thus become a strategic asset.

New considerations: When to stay away from LBA

LBA is not the key to winning every consumer. One of the common tools used in mobile marketing is a quick response (QR) code. If you have wondered whether QR codes are effective, wonder no more. A recent study in Hong Kong by Wenbo Wang investigated how the presence of a QR code on a mobile coupon affects redemptions.[14] In addition to providing insights on the effectiveness of QR codes, the study also offered more guidance on the effect of geographic location and distance on redemption. Using two large-scale, real-life experiments with restaurant coupons in Hong Kong, the author found that merely showing a QR code on a coupon increases redemption when the coupon is offered by a restaurant that is geographically close to consumers, but may backfire when the coupon is from a distant restaurant. A series of follow-up lab experiments and eye-tracking studies revealed the rationale behind this effect: Mere exposure to a QR code induces consumers to focus more on the feasibility of taking advantage of the coupon deal than on the desirability of the deal itself. Now primed for a "quick response," consumers show an even stronger preference for a coupon from a nearby store than one from a more distant store. But if the store is too far away, that delight turns into disappointment and those negative emotions can have long-term ramifications for customer satisfaction and store affinity.

This study clearly demonstrates that if firms are sending mobile offers, they should not send offers for stores located beyond a specific "optimal" distance from the target consumers. Sending coupons for "faraway" stores can backfire. You will end up priming people for a purchase, only to disappoint them later because the stores are too far away to make a "quick response" feasible. Once disappointed, they are far less likely to respond favorably to your next promotion.

New considerations: Geo-conquesting and stealing your competitors' customers

What happens if you drew the fence around a competitor, not your own store, and attempted to lure customers away aggressively? Geo-conquesting basically takes the idea behind geo-targeting and geo-fencing one aggressive step further. It means that a business will send a potential customer a coupon when that customer enters or approaches a rival store, in an effort to lure them away and win that customer.

Outback Steak House achieved some success with that tactic. For a given Outback location, a geo-fence was set up in a 10-mile radius of its casual dining competitors and a 5-mile radius around its own locations. When a potential customer entered that zone, a display ad would pop up in the Weather app or in some other location-based app. The advertisement would remind consumers of Outback's specials; clicking through would enable the consumer to see the nearest Outback location. The company claimed that top conversion activities including accessing a store locator rose by 11 percent through the geo-conquesting approach.[15] YP Marketing Solutions ran a highly geo-targeted mobile campaign for Dunkin' Donuts that targeted competitors' customers with mobile coupons[16] and reported a 3.6 percent redemption rate among mobile users who clicked and took secondary actions—almost three times higher than the average redemption rates of generic mobile coupons.

But if someone has entered a competitor's store, geo-conquesting can get very aggressive. The Guatemalan sneaker retailer Meat Pack used its mobile app to detect when a former customer entered a competitor's store.[17] As soon as that happened, that customer would receive a dynamic discount with a rapid expiration. The discount percentage would start at 100 percent and fall by 1 percentage point for every second until the consumer left the competitor's store and entered Meat Pack. That would give a consumer a little more than a minute and a half to change his mind, leave the current store, and enter Meat Pack. When the store tried this competitive geo-conquesting tactic, it managed to "hijack" more than 600 customers from its competitors. One consumer needed only 10 seconds to switch stores, which left him with a discount of 89 percent. Both Outback and Meat Pack relied on in-app messaging meaning that the customer had to have had the app running.

The retail store itself is an important battleground for advertisers trying to guide consumers' choices, especially with new technologies for geo-conquesting. The Meat Pack example goes one step beyond what Outback Steak House did, because location-based mobile advertising was used to direct consumers toward a specific business when they were physically in a competitor's location. For instance, a local delicatessen might target users who are at the businesses of more established brands like Starbucks with messages emphasizing a price promotion, a shorter line, or a new coffee blend. A Toyota dealer could serve targeted ads to customers at or near the Honda dealership a block away in a bid to convince customers to come look at its cars instead. Chrysler began to experiment with geo-conquesting in 2014, targeting visitors at other auto dealers' lots.[18] In health-care marketing, competing hospitals or clinics can capture new patients who may be looking to switch. For example, a hospital can send ads touting its short wait times to patients sitting in a clinic's waiting room.[19] The top five brand categories that are heavily adopting geo-conquesting are restaurants, retail, financial services, travel, and gas and convenience.[20]

These real-world examples are further validated by recent academic studies. A recent study by Nathan Fong, Xueming Luo, and Zheng Fang[21] demonstrates the effectiveness of *competitive* locational targeting, the practice of promoting offers to consumers near a competitor's location. The analysis is based on a randomized field experiment in which mobile promotions for a movie ticket were sent to customers at three similar movie theaters (competitive location, focal location, and neutral or benchmark locations) in a large city in Asia. Each location included a large, high-traffic outdoor shopping complex consisting of a central building housing larger merchants and separate areas housing smaller merchants. Mobile customers were offered discounts for the immediate purchase of a special offer through text messaging. The promotion was valid on the day of the offer only. The offer was pushed to customers within 200 meters of three locations. The distance between the focal and competing theaters was 2.4 miles (4 kilometers), and the benchmark location was roughly halfway between them.

The results of their experiments show that competitive location targeting can take advantage of heightened demand that a focal retailer would not otherwise capture. Deeper discounts from a competing store are effective in bringing new customers to its store. Comparable discounts from the store in which a consumer is currently located are less effective; instead of spurring sales, they cannibalize profits by encouraging existing customers to pay much less for something they probably would have purchased anyway. The good news is that marketers can use competitive locational targeting to generate incremental sales without cannibalizing profits.

There are also cost advantages. In a world without mobile phones, as store, in order to incentivize its rival's customers to come it, must display signs near its competitors or hire people to distribute flyers to make users aware of its promotions. But mobile geo-conquesting is far more efficient and cost effective.

But the study above was designed so that rival firms could not engage in this kind of poaching. If a firm's rivals were to engage in the same practice at the same time, things might well get heated. A study by Jean Pierre Dubé and some of the authors of the aforementioned study offers some guidance for marketers in such a scenario.

Suppose you have two kinds of data about your customers. For one set of people, you have their real-time location data and thus can engage in geographic targeting. For another set of people, you have behavioral data about their historical visits to your store and thus can engage in behavioral targeting. This combination of geographic and behavioral targeting can enable firms to drill down to the most relevant customers and is the most effective tactic when a firm's rivals are also likely to adopt geo-conquesting tactics.

The aforementioned study found that price competition through geo-conquesting campaigns can actually reduce the returns from geo-targeting but increase the returns from behavioral targeting![22] Competition was found to increase the profitability of behavioral targeting because firms engage in very similar pricing policies that end up softening price competition. In contrast, competition was found to lower the profitability of geographic targeting, in which firms engage in uneven pricing that end up aggravating price competition. When a single firm unilaterally bases its prices on the geographic location or the historical visit behavior of a mobile customer, it experiences a large return on investment.[23]

Interestingly, geographic targeting yields higher returns than behavioral targeting. However when both firms simultaneously engage in geo-conquesting, they experience lower returns. Thus, although competitive targeting does not result in lower profits *per se*, these researchers showed that firms will overestimate the profitability of targeted pricing if they disregard competitive responses. For firms thinking of using geo-conquesting, this is an important insight.

Maximizing Captive Shoppers: Engagement Inside a Store

Stores that have already "captured" a customer can improve their sales and profits by encouraging the customer to remain in the store longer. One could argue that the longer someone stays in a store—more precisely, the longer the path she traces—the more money she will spend. This sounds logical.

A recent study by Sam Hui, Jeffrey Inman, and their colleagues quantifies the impact of a consumer's time spent in store and the impact of mobile advertising.[24] Their study involved analyzing data on customers' paths within a grocery store in Pittsburgh using RFID tags and sending mobile offers to customers.[25] The coupons were selected from a wide range of categories at various locations in the store: toilet paper, facial tissues, paper towels, canned soup, oral care, milk and eggs, over-the-counter medicines, cereal, bottled water, flour/cake mix, cookies and crackers, pasta, and ice cream. The study found that, on average, a customer's path through a supermarket will be 1,400 feet long. A consumer's unplanned spending goes up by $1 for every additional 55 feet he or she walks in the store. That is the equivalent of about six parking spaces in the parking lot. Add mobile phone offers into the mix and the results are even stronger. The experiment revealed that mobile coupons can increase unplanned spending by 16.1 percent (in the range $13.80–$ 21.20) per customer. By strategically promoting product categories through mobile promotion with the mobile shopping app, the store could make people walk more and spend more time in the store. Walking more and spending more time lead to the spending of more money.

A follow up study by the same team of researchers deployed video cameras within stores in a city in the northwestern United States and found that unplanned purchase considerations tend to happen later in shopping trips and are most likely to occur in the aisles.[26] At the time an unplanned purchase is considered, a shopper's average mental budget for unplanned purchases is $11. Imagine the possibilities that can be enabled by mobile messages in order to capitalize on these insights!

To summarize, we now have scientific evidence that if a store can make a customer wander an additional few feet within the store and increase a customer's in-store path length, the customer is very likely to spend more on unplanned purchases. Imagine how effective a trajectory-based mobile offer could become in accentuating these kinds of behavior. In the future we will see more and more stores using beacons to instantly generate shelf-level coupons and induce additional within-store travel by offering coupons for another section of the store.

What are some of the technologies that firms can use to engage consumers once they are inside their premises? I am excited about the advent of beacons.

New considerations: Increasing indoor targeting precision via beacons

The use of beacons based on Bluetooth Low Energy (BLE) to track movements and trigger advertisements may very well turn out to be the most game-changing breakthrough in this space. The main advantage of targeting users via beacons is that it works indoors and outdoors, whereas other well-known tracking technologies such as GPS or cell tower triangulation cannot be used reliably indoors.[27] This has important behavioral implications. Smartphone users that are targeted via BLE incur lower transportation costs, since they are already in the vicinity of promoted stores. The location itself also reflects a store-specific preference, which might lead to more relevant offers. This leads, in turn, to potentially higher response rates than those achieved with more distant forms of mobile targeting, such as GPS-based geo-fences.

Until recently, users had to download a specific app belonging to an individual organization, brand, store, or event in order to get a message from it. This was often a bottleneck—many users find downloading additional apps cumbersome, as it eats up storage on their device. However, Google has recently developed a new technology, called Eddystone, whereby messages can be sent to users by triggering URLs in smartphones through the Physical Web App or the Google Chrome Widget.[28] This implies that, as of March 2016 (when Chrome 49 was released), 80 percent of global smartphones are passively beacon-enabled.[29] This new technology eliminates the need for customers to download a separate app for every individual brand, store, or event. Therefore, businesses can participate in proximity-based customer engagement without having to force customers to download a new app every time. I expect a number of new use cases to emerge as a result of Eddystone.

Dominik Molitor, Philipp Reichart, Martin Spann, and I conducted a separate study in Germany in collaboration with seventeen stores that deployed beacons.[30] Similar to what is illustrated in figure 5.2, each store

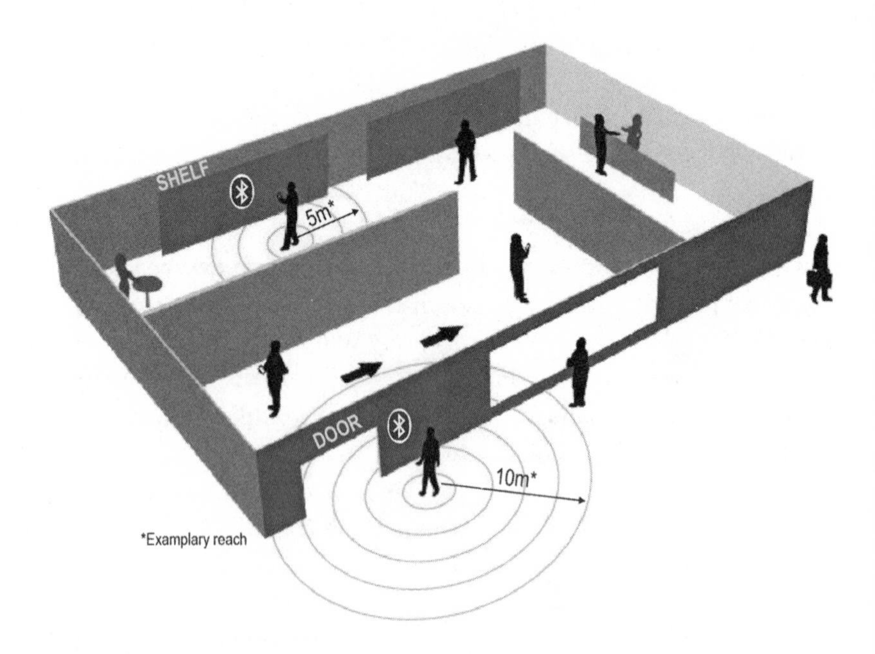

Figure 5.2
The placement of BLE beacons within a retail store.

was equipped with two beacons that acted as BLE transmitters: one at the door and one on a shelf close to the promoted product. The experiment covered eight different promoted products (two different pairs of shoes, ice cream, a beverage, an article of women's apparel, a multi-media product, a bracelet, and a cosmetic product).

We found that beacon-enabled targeted messages can increase in-store foot traffic by more than 23 percent. The content of the message makes a big difference, though. A financial incentive such as a coupon is three times more effective at bringing a customer into a store than a welcome message. Surprisingly, a retailer can also increase the time spent in store by a consumer by almost 60 percent using only simple welcome promotions, without any product-specific discount. Our results indicate that combining discount coupons and welcome promotions is an effective approach for improving retailers' revenues.

Many businesses around the world (including retail, hotels/tourism, shopping malls, entertainment, travel, real estate, banking, automotive, transportation, theme parks, conferences, health care, concerts/festivals, museums, stadiums, and airports) are already taking advantage of beacon-enabled proximity marketing.[31] Rite Aid has rolled out beacons at all of its 4,600 US stores. Other top brands using beacons include Macy's (4,000 beacons in all of its 850 stores), Target (at 1,800 US outlets), Lord & Taylor (in all of its 50 stores), and the Seattle Seahawks football team.[32] The luxury retailer Barneys uses beacons to create an omni-channel experience that involves integrating an app called The Window with data from beacons. Essentially, they send personal notifications and recommendations based on a customer's location in a store, the customer's mobile shopping bag or wish list, and the content the customer has recently viewed on The Window.[33] Besides enabling the delivery of highly targeted messages, beacons also provide insights into customers' footfalls, pathways, and trajectories, their engagement, their loyalty, and so on. From retailers in Australia to department stores in Ireland to shopping malls in the English city of Norwich to airports in Brazil, mobile-device-based proximity marketing is truly going global. Macy's is testing beacon messages outside its app and has begun to explore retargeting in collaboration with xAd.[34]

Beacon technology is also finding its way into non-profit causes.[35] Wells Fargo, sponsor of a 9/11 concert at Wrigley Field by the Zac Brown Band, used a beacon during the show to generate charitable donations to No

Barriers USA, a nonprofit dedicated to helping veterans with disabilities. The location ad platform xAd used its targeting tools on behalf of Oxfam America, an affiliate of Oxfam International, to send ads to about a million consumers at locations where they might see an Oxfam advertisement (public transportation stations, bus shelters, airports, grocery stores) and at locations such as malls and shopping centers, where people may be in a gift-giving state of mind.

Beacons can also be leveraged in the world of business-to-business marketing, Beacons can be used to send targeted information to customers within pre-defined spaces, such as a trade show. By deploying beacons and referencing the benefits of products or services in their messages, a business or an SME can use location-based targeting to attract customers to a particular booth or stand.

New considerations: Counteracting mobile showrooming

But thanks to the increasingly popular practice of mobile showrooming, you can lose sales even when your customers are in your store—right under your nose. Mobile showrooming means consumers come into a physical store to touch and feel the products before cherry-picking the best online price. Stores in which showrooming takes place can lose as much as 30 percent of their foot traffic as consumers leave the store to complete the purchase online at a competitor's store. To counter the threat of showrooming, firms have begun to use geo-fencing and geo-targeting to incentivize the consumer to complete a sale offline in their store.

A recent study by Puneet Manchanda, Stephan Daurer, Dominik Molitor, and Martin Spann has revealed some fascinating insights in the context of mobile showrooming.[36] These researchers looked at the Northern European consumer market with data being provided by one of Europe's largest and leading providers of product information and bar-code-scanning apps. The data set consisted of 80 million observations stemming from 2.5 million individual users of a product-information and bar-code-scanning app. The categories included food and beverages, groceries, media, drugstore articles, tobacco, fashion, and electronics.

It turns out that consumers search for product and price information at all locations and times, not just when they are in a store or close to one. Mobile search occurs in all stages of the purchase funnel, not just at the end. Consumers' search intensity was about 7 percent higher when stores were closed on Sundays, suggesting that potential customers carry out searches in situations that don't involve an immediate purchase decision. This study showed that showrooming is less prevalent than previously thought.

The researchers also showed that when professional ratings, consumers' reviews, and other bar-code-related information are available in the mobile app, it lowers a consumer's propensity to search for price information by about 2 percent on average. User-generated content has the biggest impact with a decrease of about 4 percent. This is a very valuable insight for stores. Including more non-price information as part of your mobile offering is a relatively easy (and inexpensive) strategy to insulate themselves against price competition.

They also found that geographically mobile consumers (those users who travel greater distances) are more active when it comes to mobile-device-based searches. This is remarkably consistent with the findings from a project that we had worked on using a data set consisting of 2.34 million mobile-device data records from 180,000 3G mobile users from a large telecom provider in South Korea.[37]

Once again, as I mention throughout the book, data science is showing the astonishing similarities in consumers' mobile phone usage behavior, irrespective of whether they are in Europe or Asia.

Food for Thought

The behavioral contradiction about spontaneity and predictability comes into play when we think about location and distance. In the simplest possible terms, we are lazy creatures of habit. Think about some of what we learned about distance in this chapter. In geo-targeted mobile advertising, the further consumers are from a store, the less likely they are to choose a related coupon. Perceived offline travel costs explain this.

What can change people's minds? Money. Consumers are willing to trade higher transportation and search costs for receiving a deeper discount in location-based advertising. That also means that firms can send a compelling mobile offer to poach on their rivals' customers in real time. All they have to do is locate consumers who are at or in a competitor's location, or have recently visited it. There is evidence that this kind of valuable advertising tactic increases local market share, foot traffic, and conversions.

A number of factors will make this straightforward relationship rather complex and therefore provide new opportunities for firms to monetize consumers'' behavior in the mobile economy. Take two stores, A and B. Store A is in the direction of a consumer's regular commute from work to home, and is two miles from her current location. Store B is one mile from her current location, but is in the opposite direction of her regular commute from work to home. In such a scenario, even though Store B is closer, the consumer will prefer to redeem the offer from Store A.

In summary, if consumers recognize their habits and routines, and if they make them visible to marketers, it encourages the ones who can go with the grain, and discourages the ones who would need to go against the grain. Prepare to enter into a new world where brands will cultivate one-on-one relationships with customers on the basis of their location histories.

Takeaways for Firms

By 2017, location-specific mobile ads are expected to account for about 55 percent of all mobile ad spending.[38] Here are some takeaways.

• Be prepared to make geo-targeting, geo-fencing, and geo-conquesting elements of your overall marketing mix. Keep in mind that price competition through geo-conquesting can reduce the returns from geo-targeting but increase the returns from behavioral targeting.
• A business can literally use money to shorten the distance between itself and a consumer. This can be done either via a mobile push or a mobile pull promotion.
• Be mindful of how far you are from the customer when you send the offer. Sending mobile offers to customers located beyond a certain distance or to customers who have to change their usual direction of travel to redeem the offer, can backfire because that people may get primed for purchase from the initial offer only to be disappointed subsequently.
• Beacons are an excellent tool, both for luring in consumers and for micro-location targeting once consumers are inside a store. They can be used for messaging to a captive audience in any venue. This can be deployed both in the business-to-consumer)world and the business-to-business world.
• Firms can alleviate price competition from mobile showrooming by providing user reviews and professional ratings in their dedicated mobile apps.

6 Time: It's On Your Side

The statement below, uttered by a police officer in a television show, seems to capture the zeitgeist of the 21st century perfectly:

> I don't know, maybe part of it's the fact that you're in a hurry. You've grown up on instant entertainment. Instant communication. Instant transportation. Flash a card—instant money. Shove in a problem—push a few buttons—instant answers. But some problems you can't get quick answers for, no matter how much you want them.

If you think that sounds like just another rant against the Millennial generation, which takes the wonders of the smartphone for granted, you should think again. The target of the rant was a group of Baby Boomers. Those words came from Sergeant Joe Friday, hero of the television series *Dragnet,* in an episode that was first aired on March 7, 1968.

We can interpret that as meaning that some things never change. Older generations will always be critical of younger generations. What has changed, however, is our perception of what the word "instant" means.

Compare these two situations, both involving the instant major news broke during a national sports broadcast. On the night of December 8, 1980, viewers of ABC's Monday Night Football became the first to hear of the death of John Lennon when the commentator Howard Cosell broke the news during the telecast. No other news outlet had the story. ABC found out about Lennon's death by pure coincidence when one of its producers happened to be an emergency room patient at the same hospital where Lennon was taken after the shooting.[1] Other television networks picked up the news after ABC's announcement. But if you were at the game, or if you weren't watching television that night, you probably didn't hear about the incident until the next morning.

In May 2011, a similar situation unfolded during ESPN's national baseball game of the week. The 40,000 fans gathered in Philadelphia's Citizens Bank Park inexplicably began chanting "U-S-A, U-S-A" with no apparent prompt or reason. A video clip of the moment shows the players and umpires on the field looking confused and shrugging their shoulders.

As the chant grew louder, ESPN play-by-play announcer Dan Shulman cleared up the mystery for the TV audience: US Special Forces had killed Osama bin Laden. How did the fans find out? There was no news displayed on the scoreboard and no news announcement over the public address system. But the fans had smartphones! The news spread swiftly and clearly enough to bring that spontaneous patriotic chant to life.[2]

"Instant" continues to take on more immediacy, and leads to questions about the importance and relevance of time in an era in which 84 percent of smartphone owners in major US cities check an app first thing in the morning.[3] How can advertisers use time to their advantage to reach their target audience? Now that smartphones give advertisers the opportunity to reach an individual consumer in real time, what is the secret to getting the timing and the ad right? What is the best way to deliver a timely advertisement that will trigger a positive response?

To answer such questions, we first need to be clear about what we mean by "time." The obvious answer is "time of day," and it is reasonable to think that consumers respond differently to ads depending on the time of day. But we also need to consider the day of the week. How does consumer's behavior change, if at all, on different days of the week or on weekends? "Time" can also mean "duration." Does our sensitivity to advertisements change if the expiration date of a coupon or offer is closer or farther away?

My colleagues and I have conducted several studies to answer those questions in ways that can help mobile advertisers approach their "instant" opportunities with greater savvy and confidence.

The Advent of Granular Advertising

The rate of ad creation has increased steadily since World War II, in lockstep with the rate of content creation. At the same time, the complexity of decisions about the best ways to take advantage of that speed has also increased.

Through the 1940s, the daily newspaper was the cornerstone of a consumer's morning, together with the radio. Average daily paid newspaper circulation in the United States stood at 53.8 million in 1950, corresponding to 123.6 percent of households. For the thicker Sunday paper, laden with advertisements, paid circulation was 46.6 million, or 107 percent of households.[4] By 2010, paid circulation for daily papers had dwindled to 43.4 million and to 44.1 million for Sunday papers. That may not seem so dramatic, until you compare the share of households. Circulation in 2010 represented only 36.7 percent of households (daily) and 37.3 percent (Sunday), representing declines of 70 percent and 65 percent, respectively. Paid circulation for newspapers in the United States now hovers at just over 40 million.[5]

Television started the decline of physical newspaper circulation, and in many ways, the Internet accelerated its decline. With this dramatic shift in media consumption came a step change in how advertisers could target consumers as a function of time. In the era when newspapers and periodicals dominated media, consumers received a fixed batch of ads daily, weekly, or monthly. Advertisers also needed to take decisions on content and placement well in advance. Immediacy—as we perceive it today—did not and could not factor into their decision making.

Television and radio gave rise to the practice of "dayparting," which means dividing up the consumers' day into segments, each with its own particular audience and content. We are all familiar with expressions that define these segments, such as "rush hour," "prime time," "drive time," and "daytime." There is an underlying assumption that the audiences in each segment are relatively homogeneous. These ads run, however, independently of whether the device is on or how much attention the viewer or listener is paying. Agencies such as Nielsen and Arbitron (before their merger) estimated the size of audiences, but it was impossible to get precise data on how many people were exposed to an advertisement, who these people were, or anything beyond a rough aggregated level.

The Internet created a new dynamic by allowing consumers to trigger an exposure to an advertisement on their own. For the first time, advertisers could collect data at the individual level on the time of exposure. Granted they have to deal with the "viewability challenge": who actually viewed the ad as opposed to an user being served an ad that appears in an inactive Web window or an out-of-view part of their screen. The advent of ad blockers makes viewability a challenge. In addition, the fact that mobile ads can take up to 5 seconds to load, on average, aggravates the problem because slow ads can also drag down the page load time of a site causing users to exit the site.[6] Recent ad tech companies have developed viewability tracking tools to learn who has viewed the advertisement, and in some cases, how he or she behaved after exposure to the advertisement.[7]

The smartphone took this trend one step further. The mobile Internet created the first ubiquitous, universal advertising channel that enabled individualized 24/7 targeting. Because consumers rarely stray far from their smartphones, marketers can now actively select the preferred time to advertise their products, either through the app, through the mobile browser, through a text message, or through a push notification.

For marketers these possibilities may be exciting or paralyzing. When the primary basis for decisions was experience, estimates, intuition, and best judgment, no one could demand precision, because no one could expect it or even measure it. There were no means to get better data in order to legitimately challenge or overturn an industry rule of thumb.

Today when you can target individuals at any time of day and be reasonably sure of reaching them, all of a sudden you can expect and demand precision. You should be able to find when jackpots and dead spots are most likely to occur in a consumer's day, and surgically target consumers with customized advertisements.

Nonetheless, until recently there was a serious lack of consensus on when these time periods are or whether they even exist. Some recommend eschewing harder analysis and distributing mobile ads evenly across the day to cover all variation in consumers' tastes and interests. Because our earlier understanding was based on far more aggregated information, it yielded far less granular insights. Some claim that midday is the best time to serve an ad,[8] others cite the evening,[9] and still others highlight several peak periods between 9 a.m. and 3 p.m.[10] But the reality is that most people experience many different micro-moments (there is that Google concept again!) between 9 a.m. and 3 p.m. Hence, it makes sense to think of customized advertising based on the time of the day, or based on these micro-moments.

New considerations: Advertising effectiveness based on time of day

The concept of micro-moments requires more precise guidance on the effect of advertising on different hours of the day vs. the conventional guidance that broke out the day in big chunks of time. To address this, a team of researchers in the United States and China conducted two large-scale, real-life and very creative studies. In the first study,[11] they divided the day up into twelve hour-long intervals, each beginning on the hour, from 8 a.m. to 7 p.m. They assigned two groups of people to each interval, for a total of 24 groups. In each time interval, one group would receive a promotion for a high-involvement practical (utilitarian) product and the other a promotion for a lower involvement, more indulgent (hedonic) product.

The response rate for the test was 1.94 percent, more than four times the response rate of 0.42 percent for mobile targeting in general.[12] The test revealed that response rates for mobile ads for utilitarian products are highest in the morning, modest at noon, higher in the afternoon, and low during the evening. In contrast, response rates for hedonic product mobile ads are low in the morning, highest at noon and afternoon, and modest in the evening.

In detail, utilitarian products show two pronounced peaks relative to their weakest period (7–8 p.m.). Sending mobiles ads between 10 a.m. and noon increased the likelihood of a purchase significantly. The response rates for hedonic products showed similar peaks, but at different time periods. In their case, the base (weakest) interval is 8–9 a.m. Response rates for ads sent between noon and 2 p.m. boosted purchase likelihood by a factor of 7.1 on average, with other variables held constant. The simplest explanation for these very sharp differences is that consumers tend to be more businesslike in the morning hours, but then relax progressively toward the afternoon.

As insightful as this experiment was in pointing out where the jackpots are for ad responsiveness by product type, it raised another question: what are the effects if the consumers see an ad for the *same* product, and the only difference is whether that product is described in utilitarian "work" way or a hedonic "play" way? If the design and phrasing of the value proposition makes a difference, it has implications for the creative side of advertising, not only for its timing.

To explore that, this team conducted a similar test. Across the twelve time intervals, consumers saw an ad for one product with either a utilitarian or a hedonic framing. An example of a utilitarian framing from a telecom provider selling high-speed Internet access would read "Download big email attachments in the blink of an eye and handle official business anywhere you go." In contrast, the hedonic framing would read "Feel free to download undamaged music, watch HD video, and never get blocked playing games anywhere you go."

The results of the experiment echo a comment made by the advertising pioneer Leo Burnett: "There is an inherent drama in every product. Our No.1 job is to dig for it and capitalize on it."[13] Framing an advertisement for a product in a utilitarian way reduces the purchase likelihood by 80 percent compared with a hedonic framing for the exact same product. However, there are still peaks and valleys, which follow patterns somewhat similar to what we saw in the first experiment.

For the "work" or utilitarian framing, sending an ad between 10 a.m. and noon increased purchase likelihood by a factor of roughly 5 relative to the base (weakest) period of 7–8 p.m. For a "play" or hedonic framing, sending a mobile ad between 1 p.m. and 3 p.m. increases purchase likelihood by a factor of 2.4 relative to the base (weakest) period, 8 –9 a.m. The size of the differences for the "play" value proposition is smaller than for the "work" ones, but the base is higher for the "play" ones. This explains why in general the "play" framing is the better choice, assuming the advertiser prefers only one option.

The key takeaway from these results is that micro-moments in the afternoon for functional products can be handled with hedonic framing of the message. Similarly, micro-moments in the morning for hedonic products can be handled with functional framing of the message. By simply changing the framing of an ad for the same product, managers can take advantage of more micro-moments through the day.

These studies provide us with precise measures of the true effectiveness of time-based mobile advertising and enable us to predict consumers' responses at a level of accuracy that was unfathomable even a few years ago. The qualitative insights may not be shocking, but the quantitative insights can be very useful.

New considerations: Advertising effectiveness based on day of the week

Does the day of the week change how we respond to advertisements? Yes, it does. A study led by Peter Danaher found significant day of the week effects. Mondays and Thursdays have significantly higher redemption rates than Wednesdays. According to the researchers, the Monday and Thursday effects can be explained by the fact that weekday visitors to the mall are likely to include a high proportion of people who are not employed full time.

My own studies conducted with Beibei Li and Siyuan Liu have also found very significant day of the week effects. At a high level, we can divide shoppers into two categories: focused shoppers and explorers. Consumers are more likely to be focused shoppers during the week, meaning we have a mission or purpose in mind. This increases their sensitivity for advertisements relevant for our mission, but prompts them to ignore and even resist the less relevant ones. On the weekends, consumers tend to be explorers rather than focused shoppers. In that case, broadly speaking, the effects are reversed. They are much more receptive to random advertisements on weekends.

Previous marketing and psychology literature back up these findings. When customers are in the unplanned or "explore" stage of the purchase cycle, they are more likely to make impulse purchases.[14] A consumer's propensity to purchase on impulse gets an additional boost when he or she sees a random item on sale.[15]

Companies are also experimenting alongside academics to better understand the effectiveness of advertising on different hours of the day. Taco Bell is using the real-time mobile location intelligence platform from xAd (known as Marketplace Discovery) to uncover time-of-day trends as well as day-of-week trends, and is harnessing these data to make more informed marketing decisions and enhance the effectiveness of its limited-time promotions.[16]

New considerations: Effectiveness of coupon redemption windows

Now let's look at time from another perspective—the length of validity of the offer. How long a consumer has to make a purchase decision is also an important element in designing a mobile marketing offer. Previous tests have verified that the timing of promotions impacts their effectiveness.[17] Some consumers are also prone to spend when they receive an in-store promotion for an immediate purchase.[18] In a study[19] (led by Peter Danaher and his colleagues) of several stores in a shopping mall in a large Western country, the most represented product type was snack food. In addition, the study included several other product types, including clothing, shoes, electronics, hair-styling products, and novelty gifts. The researchers found that the validity period of mobile coupons is an important determinant of redemption rates. This happens because redemption times for mobile coupons are typically much shorter than for traditional coupons.

The researchers also found that once users receive the offer, the time to redemption is very quick, with a median redemption time of less than 2 minutes. For traditional grocery store coupons redemption rates are highest just after a coupon is dispatched and just before the coupon expires.[20] In contrast with traditional coupons, this team found no evidence that mobile coupons are stored to be redeemed later and no increase in redemption as a coupon is about to expire. In fact, the maximum time to redemption was only 16 hours. This is remarkable. The average redemption window for mobile coupons in their study was nearly ten days, and yet all coupon redemptions happened in less than a day. This implies that mobile coupons should have much shorter redemption windows than traditional coupons. The researchers quantified that if redemption windows were to be shortened to one day for all coupons, the predicted redemption rate would increase by 50 percent over the current rate!

A key managerial insight from their study is that when the context and location are appropriate (for example, a shopping mall), advertisers in many product categories can maximize redemption rates by shortening the redemption window in order to signal time urgency. As the authors suggest, messages such as "limited to the first 100 customers today" or "only available until 2 p.m. today," in the coupons can accomplish this goal. Embedding such messages mobile offers would evoke the existing use-it-or-lose-it mentality of traditional coupons while keeping the expiry length more appropriate for mobile coupons.

One issue that many firms have pondered about (based on my discussions with executives in various industries) is whether mobile marketing only generates immediate sales as opposed to a more delayed or lagged sales, at least occasionally. For example, is it possible that a mobile promotion leads to a sale after a couple of weeks as opposed to having an immediate redemption? Social psychologists have taught us that in order to activate immediate purchase behavior two things must happen. First, firms must trigger a sudden and unplanned consumption urge. Second, firms must create a psychological state that allows the desire to instantly fulfill the consumption needs by outweighing potential inhibiting factors.[21] Since mobile promotions can be designed to create location and time congruence, it is easy to imagine why the mobile channel to activate both elements in a consumer's mind.

But what about delayed sales impact? Can mobile marketing create a long-term impact? Yes! According to the theory of planned behavior that Philip Kotler describes in his book *Marketing Management*, a consumer makes a purchase in five stages: problem recognition, information search, evaluation of product options, purchase decision, and post-purchase support.[22] Upon being exposed to a mobile promotion, the various features of today's smartphones can take a consumer through each of these five stages. Promotion messages can be stored in a mobile phone, users can solicit opinions from friends and family, users can coordinate with others for a joint consumption experience, and so on.

To investigate this matter in a data-driven manner, Zheng Fang, Bin Gu, Xueming Luo, and Yunjie Xu ran a number of interesting field experiments in the summer of 2014, using 22,000 participants and multiple control groups. Some participants received location (geo-fenced) ads; other received non-geo-fenced mobile ads.[23] The team found that location-based mobile ads can have significant delayed sales effect for up to 12 days after the mobile promotions. Importantly for brands, the total sales impact of mobile offers could be underestimated by 54 percent if marketers did not include the delayed sales impact and considered only contemporaneous impact.

The study by Fang et al. goes a long way toward conclusively establishing that mobile ads or mobile promotions can have a delayed effect several days after the initial exposure.

New considerations: Exploring the relationship between redemption windows and geography

An important issue for consumers seems to be the combination of duration and location. My colleagues and I have demonstrated that at one extreme—if we are not close to the store that makes us an offer via mobile advertisement, and we have no sense of urgency—the offer is not very compelling.[24] This seems rather intuitive. The perceived benefit of receiving promotions for events occurring far away from us and further in the future is low. We find such events harder to visualize in a concrete way.[25]

But what happens as we move away from that extreme? In other words, how does our responsiveness change if the events move closer to us, both in terms of distance and timing? The first step is to test what happens at the other extreme. To do this, Michelle Andrews, Xueming Luo, Zheng Fang, and Chee Wei Phang observed in a sample of moviegoers whether people could be influenced to watch a movie at a nearby IMAX theater, depending on how much advance notice they received.[26] The wireless provider sent text messages promoting discounted tickets to movies. Recipients purchased movie tickets by downloading the accompanying movie ticket application and ordering from the app. For the same-day effect, they sent some users a coupon at 2 p.m. on a Saturday (two hours' notice). For the one-day effect, they sent a similar coupon at 2 p.m. on the preceding Friday (26 hours earlier), and for the two-day effect they sent the coupon at 2 p.m. on the preceding Thursday (50 hours earlier). All coupons were offering the same discount: 50 percent. They also exploited proximity: the distance between the theater and the smartphone users in the experiment when the advertisement was sent ranged from 200 meters to 2 kilometers. Three groups were identified: proximal users (within 200 meters), medium-distance users (between 200 and 500 meters), and faraway users (between 500 meters and 2 kilometers). They sent text messages to mobile users in the area to promote the movie. The response rate was 7.35 percent.

What did the study find?

The results of the timing were significant and unambiguous. The longer the lead time (coupon expiration time), the less likely someone was to go the movie. Same day (two hours) notice outperformed one-day notice, which in turn outperformed two-day notice. These findings are in line with the results of previous studies and support the idea that in the 21st century real-time marketing brings results and is worth the investment.[27]

To further test the combination of urgency and distance, Andrews et al. extended the initial field-experiment-based study and collaborated with a market-research firm to conduct a survey. Respondents were asked to comment on one of these four scenarios:

- scenario 1: close to the theater, short notice
- scenario 2: close to the theater, longer advance notice
- scenario 3: farther away from the theater, short notice
- scenario 4: farther away from the theater, longer advance notice

The survey asked the respondent to imagine that he or she was a specified distance from an IMAX theater and received a text message with a promotion for the 4 p.m. Saturday show. The respondents in scenario 1 (close, urgent) showed the highest level of involvement and intent to purchase—significantly higher than the consumers in scenario 2. In detail, nearby consumers were 76 percent more likely to attend the show when they received same-day notice than when they received two-day notice, with other variables held constant.

A surprising finding

The worst-performing scenario was scenario 4, which supports the old cliché "out of sight, out of mind." Participants in that scenario had the most difficulty seeing themselves taking advantage of the offer. Participants in scenario 3 also found the offer to be intrusive, which highlights the interplay between greater distance and short notice.

The revelation in this experiment came when one took a closer look at scenarios 3 and 4. The above findings certainly held up for people who are close to or a medium distance away from the theater. However, for people who were far away from the theater, neither too little lead time (same day) nor too much lead time (two days) would deliver the most benefits; in fact, redemption rates were highest for those who received offers one day before the screening. The optimal amount of advance notice was one day. Same-day notice was too abrupt for the consumers to respond, because of the distance and the associated hassles of travel. Two-day notice was also not optimal, because the longer lead time created too much uncertainty. One-day notice was ideal for consumers to envision attending the movie and to plan accordingly. Giving the farther-away consumers one day of notice increased their chances of attending by a factor of 9.5 vs. same-day notice, and by 71 percent vs. two-day notice, with all other variables held constant.

In summary, geographic targeting (delivering an ad at the right place) and temporal targeting (delivering an ad at the right time) on smartphones are both very effective. But when we combine the two, the results are fascinating and often surprising. For marketers targeting mobile customers located close to them, sending an advertisement that expires the same day is better than sending an ad with a longer lead time. For marketers targeting mobile customers located farther away, sending an advertisement one day before is optimal. If marketers give too much or too little lead time on a promotion, the targeting effectiveness plummets. With too little time, consumers may find it difficult to reach the event on time, and will not have an opportunity to plan to take advantage of the offer. Too much lead time, in contrast, reduces their involvement. They are less likely to consider the offer, and those who do are less likely to purchase.

What is the psychological theory explaining these results? According to Andrews et al., construal-level theory explains how consumers differentially evaluate mobile messages under varying contexts of location and temporal conditions, which leads to changes in effectiveness of mobile targeting.[28] Construal-level theory tells us that consumers' purchase intentions are highest when they receive a text message close to the time and place of the promoted event. Andrews et al. assert that this occurs because shorter temporal and geographical distances induce consumers to mentally construe the promotional offer more concretely, which, in turn, increases their involvement and purchase intent.

Not all of these insights will translate into every other product category. For more elastic products, changes in discount percentages may have a disproportional effect on purchases among customers. The length of the product life cycle will influence the effectiveness of temporal targeting strategies. Going to the movies is a social event and requires time-consuming coordination. Movie tickets are relatively inexpensive and therefore more likely to be subject to impulsive purchase, peer pressure, and word-of-mouth influence.

Nonetheless, because the interaction of time and location makes contextualized marketing more complicated than when firms think of each force separately, the study by Andrews et al. provides firms an excellent benchmark for balancing temporal and geographical mobile targeting strategies. As the authors themselves assert, "treating time and location separately in a 'one plus one' fashion will not work," and "managers must carefully design mobile campaigns that balance the goals of timely information against the risks of alienating customers through a shotgun approach."

New considerations: Mobile advertising and its effect on different stages of the purchase funnel

Many of the studies discussed in this section highlight the importance of the absolute timing of an exposure to advertising (time of day, day of week, etc.) on the advertising's effectiveness. However, one can also define the timing of the exposure in a relative sense—for instance, timing of an advertising exposure can be defined as a function of where a consumer is in the "purchase funnel." One of the classical marketing models that academics and practitioners have adopted is the AIDA framework,[29] which identifies the cognitive stages an individual goes through during the buying process for a product or service. According to the AIDA framework, there are different stages in a consumer's path to purchase: awareness, interest, desire, and action. One would expect consumers' responsiveness to advertising to vary substantially as they progress through those stages.

In a recent study that the dynamics of effectiveness and tradeoffs in digital advertising, Param Vir Singh, Vilma Todri, and I analyzed consumers' responsiveness to digital advertising as they progress through the latent stages of the purchase funnel.[30] Using a data set from a large US-based retailer, we investigated the enduring impact of display advertising on various aspects of consumers' decisions during their path to purchase. We were primarily interested in examining the potential of display advertising to trigger annoyance in consumers as a function of the relative timing of the advertising exposure (i.e., stage of the purchase funnel that consumers are currently in). We were fortunate to have had access to highly detailed information about the individual-level viewability of display advertising exposures.

Upon analyzing our data, we found concrete quantifiable evidence about the tension that brands face between trying to generate interest without triggering annoyance as consumers move along the funnel path. While display advertising can have an enduring impact on transitioning consumers farther down in the funnel path, we showed that exposures beyond a certain number within the same day substantially increases the probability of triggering annoyance. Although this reaction is expected, our study focused on understanding annoyance thresholds at different stages of the purchase funnel. The study validated that consumers who reside in different stages of the AIDA funnel path exhibit substantially different thresholds for annoyance stimulation. For example, consumers who are in the "awareness" stage can be annoyed with as few as three display advertising exposures on the same day. In contrast, consumers who are in the "interest" state would have to be exposed to seven display advertising exposures on the same day to reach a state of annoyance. In the same vein, by strategically varying the timing and number of exposures, display ads can also be used to influence people to make a purchase. Two exposures per day is the optimal number for consumers who are in the "awareness" state and four exposures per day is the optimal number for consumers who are currently in the "interest" state.

As consumers use multiple platforms and engage differently with brands at different touchpoints in their path to purchase, firms need to be cognizant that the timing of their mobile ad exposure has a dramatic effect on consumers' behavior. It is extremely important for businesses to keep in mind that how and when they communicate with a consumer on his or her mobile device should also be a function of where in the purchase funnel that specific consumer is.

Food for Thought

The Carousel of Progress at Disney's Magic Kingdom traces the evolution of labor-saving devices. In a handful of stages representing periods from the beginning of the 1900s to the 21st century, the audience sees how we get more and better leisure time as the range and quality of appliances improve.

I sometimes think of mobile advertising as the digital version of a labor-saving device, and we are only at the beginning of its evolution.

Think back to the behavioral contradiction about advertising's being a nuisance unless it helps us save time and avoid missing out on exciting opportunities. If marketers can use the data-driven insights about time and location to deliver on the promise of "the right message at the right place at the right time," the social and economic effects for consumers and marketers alike can be dramatic. Shopping becomes more efficient. Awareness rises, search costs fall, and consumers can plan better and invest their free time elsewhere.

Of course, this will require some effort on the part of marketers. When businesses start to weave in a modest amount of additional information—such as product category and location or distance—they get a sense for how these behavioral drivers work in combination. The best way to target a consumer at the "right place at the right time" is now something a firm can figure out with confidence for an individual.

Thus, I recommend that consumers ask themselves "How much is my time worth?" Can you shave even more time from that labor-intensive activity called shopping that eventually drives your lifestyle? It took us almost a generation to take online commerce for granted. Mobile commerce is the next step. If you, the consumer, let marketers learn your habits and read your mind, you will let that smartphone become your concierge and save you some more time.

Takeaways for firms

• The mantra "right message at the right place at the right time" is no longer a dream. It should be a day-to-day reality you work on perfecting, even if that requires some upfront investment. Time has many dimensions, all of which influence our behaviors and advertising effectiveness in the context of the mobile economy. They range from the current "real" time to the stage of the path to purchase funnel to the period of the day (e.g., rush hour), to the day of week (weekday or weekend) or even length of the validity of a mobile offer.

• For utilitarian (or functional) products, mobile purchase rates are highest in the morning. For hedonic products, mobile purchase rates are highest in the afternoon. However, for a given product, micro-moments in the afternoon can be made more effective with a hedonic framing and micro-moments in the morning can be made more effective with an utilitarian framing. By changing the framing of an ad for the same product, firms can take advantage of micro-moments throughout the day.

• Think about whether your goal is to ensure a planned and delayed purchase or trigger an immediate and contemporaneous purchase. You can achieve that goal by offering a very short or a prolonged window of redemption in your mobile promotion.[31] Shortening the redemption window of a coupon is a powerful instrument to increase its effectiveness, as long as the context and location are accounted for. That said, mobile ads have been documented to have had a delayed sales effect of up to 12 days.

• Geographic targeting can be combined with temporal targeting to create powerful synergies towards increasing redemptions of mobile offers. But too much or too little lead time can be detrimental for offer redemptions.

• The format, timing, and the frequency with which a business communicates with a consumer on his or her mobile device should be a function of which stage (awareness, interest, desire, or action) in the path to purchase funnel that specific consumer is at.

7 Saliency: Can You See Me Now?

In the age of smartphones, it is hard for us to imagine how limited and rudimentary our access to information used to be. Think of a family traveling by car from one city to another. Evening approaches and the parents decide it is time to stop for the night. They think "Where should we eat? Where can we find a hotel or motel? What else can we do around here?"

Answering those questions was a much different experience in the 1980s than it is today. Back then, one Massachusetts county operated a tourist information center at a rest area to provide travelers with such services. Located along Interstate 95 just north of the border with Rhode Island, the center offered a selection of leaflets and brochures on attractions, food, and lodging in the Greater Boston area. It also had a help desk where advice could be provided. After sizing up the traveling party—size and mix, interests, available time, willingness to pay—the person at the help desk would make concrete recommendations until the travelers made a choice, and would even make tour, dinner, or hotel/motel reservations if the travelers desired.

Unbeknownst to the travelers, the offers the person at the help desk made were not arbitrary. Various hotels, restaurants, and hotel operators paid advertising or sponsorship fees in return for getting more prominent placement or mentions. Some hotels set aside small contingents of discounted rooms to try to lure travelers all the way into Boston.

A typical conversation went like this:

Staff: "Well, if you want to stop now there is a place at the next exit where you could be looking at $59 for the night. But if you go an extra 10 miles up the road, there is a local motel—not a chain—which has rooms starting at $29."

Family: "And what if we drive all the way into Boston?"

Staff: "Then we can get you right downtown for $85 a night if you stay two nights"

It was a very primitive and subconscious form of search engine optimization, complete with all the biases and pseudo-precision humans are prone to inject into what amounts to an ad-hoc process. If the family filled out a questionnaire afterwards, the tourist information center's team may have just enough data to draw the small-sample conclusions we are always tempted to project to full-fledged reality, especially when they support one's current thinking.

Today, a family might never stop at the rest area unless they need to take a "bio break." "Tourist information" means asking mom, dad, or the 14-year-old in the back seat to pull out the smartphone (assuming it isn't already in their hands), then do a quick search on Google or in their app of choice. They would pull up an instant and much longer list of attractions, hotels, and food options. Depending on the degree to which the family has opted in and made its data history available, that list would be customized to them and their needs.

Intuitively we would expect the similar mental algorithms from the 1980s to apply to the data-driven ones today. The staff member at that tourist information center was onto something. Intuitively, the customer is liable to choose an earlier option than a latter one (in the order they are presented), and those options further down the list had better be very good deals to prompt the customer to act.

This finding may be intuitive, but it has an intriguing psychological basis. In our efforts as consumers to find our "best right answer" to a question, we follow a certain pattern when we recall what we have seen and make our choice. We are more likely to recall the items at the beginning of a list (primacy effect) and at the end of the list (recency effect) than the ones in the middle. The first item on a list may not objectively be the "best right answer," but it has much better odds of getting selected. It is the easiest to click, because it requires the least search activity.

This brings us back to two of the behavioral observations made in part I: We find advertising annoying, but we fear missing out and we would prefer not to waste time in the trial-and-error process of searching for what we need. We want choice and freedom, but we also get easily overwhelmed. Imagine an ideal world where the "best right answer" jumps out at us immediately, even if a search returns hundreds or even thousands of viable "right answers." We don't need to scroll down and squint to find what we want. We don't need to refine and repeat our search. We don't need to make a tough call. We always get the "best right answer" with the least possible effort.

We refer to this as *saliency* or the *position effect.* Consumers want to see the best right answer stand out on their screens. Advertisers, retailers, and other marketers want their message to be that "best right answer."

Does Ranking Matter?

Ranking is a great lens for firms to view how well saliency is working. Depending on the platform, ranking might be a function of distance, relevance, the result of an auction process, or randomized. Rankings play an important role in platforms such as Yelp and TripAdvisor, which aggregate listings and display them using different relevance-based algorithms. Auction-based rankings play a critical role for marketers in search engines. In the world of mobile-based retail apps, our research has shown that distance between the user and the store plays an extremely important role in influencing purchases. Ranking is sometimes perceived as a proxy for distance.

Over the last decade, my work with Sha Yang on sponsored search advertising[1] has documented that standing out in a search engine makes a real, positive financial difference. Highly ranked advertisers achieve a concrete financial benefit from better saliency, because of the interplay between top-ranked search results and a greater number of clicks, conversions, and revenues. Outside of the world of search engine advertising, my colleagues and I have seen similar ranking effects in travel search engines like Travelocity and TripAdvisor too.[2] Tweaking the algorithms that facilitates ranking on these platforms has enormous implications for hotels, platforms and consumers. In our work, we had shown that by carefully harnessing user-generated content and crowdsourced content from the Internet, travel search engines can significantly alter how their ranking systems affect consumers' choices.

I have also witnessed similar consumer behavior in the field of social media. In chapter 5 I described a study in which my colleagues and I looked at the impressions and engagements for a microblogging service. The closer a post appeared to the top of the screen, the greater the likelihood that a user would click on it. For users of personal computers, an improvement of one rank increased the likelihood of clicking by 25 percent. Another study of the retail environment (led by me) suggests similar findings. Scrolling down one ranking level reduces response rates by approximately 5 percent depending on the category of products advertised. Similar consumer behavior has been witnessed in the field of email marketing too. More than a decade ago, Asim Ansari and Carl Mela showed a positive relationship between the serial position of a link in an e-mail (the equivalent of its search ranking) and recipients' clicks on that link.[3]

These studies demonstrate unequivocally that there is a strong psychological ranking effect on people's choices. And that is why search ranking matters and why Google has been able to build an ad business worth more than $70 billion on that premise.

Ranking is likely to become an important design option for aggregating content and displaying advertisements in mobile applications for Yelp, Foursquare, TripAdvisor, Groupon, and other platforms. This is especially true for Yelp, which benefits primarily from greater salience in "local" search. Local search now accounts for more than half of all searches on Google for many of these platforms. More Google searches now take place on mobile devices than on computers in ten countries, including the United States and Japan.[4] Searches with local intent are more likely to lead to store visits and sales within a day. A Google study showed that 50 percent of mobile users are "most likely" to visit after conducting a local search, versus only 34 percent of consumers who searched on tablets or computers. Roughly 18 percent of local searches lead to sales, versus 7 percent for nonlocal searches. Given how important mobile phones have become to their search business, Google made a big move on November 4, 2016. Instead of using the desktop version of a page's content to evaluate its relevance to the user, Google decided to change its algorithm to give more importance to mobile indexing. From now on, its algorithms will primarily use the mobile version of a site's content to rank pages from that site, and to show snippets from those pages in its search engine results.[5]

What can firms do to improve their ranking in local search? They can promote the quality and quantity of user-generated content. Google's search engine optimization algorithm gives a lot of importance to user-generated reviews on Yelp. Previously, Google manipulated its algorithm to show its own local listings and content above Yelp pages, even if a query included the word "Yelp." Google now takes into account the presence of a business on Yelp, such as the number of reviews it has and how positive they are.[6] When you query for a restaurant review on Google, the first few hits on a search engine are often reviews from Yelp. This better ranking gets Yelp a lot of consumer traffic.

Saliency and ranking are huge consideration in the B2B world too. For a business to be able to do business with other businesses, it has to be discovered on a mobile platform.

So let's focus on ranking in this chapter, how it relates to saliency and to other forces that shape the mobile ecosystem. Considering the current ubiquity of smartphones and the expected emergence of all sorts of information-enabled devices, these intuitive insights raise some questions whose answers could be very lucrative for both the app ecosystem and the businesses it serves.

New considerations: The saliency tax and its drivers

We know that searching and scrolling on a smartphone presumably requires more effort than on a personal computer or a tablet, which we as consumers want to minimize. We want our "best right answer," and we want it while incurring the least amount of search cost. My colleagues and I set out to quantify this burden. Think of it as a "saliency tax."

In the era of smartphones, we see a physiological effect in addition to the psychological ones. The primacy effect in particular is intensified by the small screen, which displays the search results the consumer needs to review. The average consumer can take in hundreds of products at a glance through their direct and peripheral visual fields when they enter a supermarket or a warehouse store. The personal computer already narrows the field of vision considerably. But the smartphone limits the field to a battlefield of between roughly 10 and 15 square inches in size. This amplifies the importance of ranking when a consumer looks at search results.

My studies[7] have demonstrated that the ranking effect is stronger—and thus the search costs are higher—for mobile users than for users of personal computers, presumably because of the differences in screen size. For the user of a personal computer, an improvement of one rank increased the likelihood of clicking by 25 percent, while the same improvement increased the odds of a click by 37 percent in the case of mobile users. Another way to think of this finding is that the odds of someone clicking are 6 percent higher on a mobile device than on a personal computer with the improvement of one rank. Yet if we looked at the 5th highest search result, the odds of clicking are 26 percent *lower* on the mobile device than on a personal computer. When we go down to the 10th highest search result, the decay continues. The odds of a click from a user of a mobile device are 52 percent lower than for a user of a personal computer. The cumulative effect of such differences again underscores the premium value to appearing among the initial results provided by an app or a search engine.

Scrolling on small screen takes effort and concentration. As was noted in part I, effort and concentration are in short supply when people are looking for their "best right answer." It is like walking into a department store wearing blinders with lenses that make the print and the images smaller. To get a feel for this, cup your hands around your eyes and then look out of a window with a panoramic view. You will quickly notice two effects: you focus more intensely on what you see in your limited field of view, and as you look around it is hard for you to put together a complete overview of what you are looking at. (You'll get an appreciation for that if you now uncup your hands and take in the full panorama.)

Why does the small screen increase effort and require more concentration? This is known as *information chunking*. It means that we take an intense look at what we see in our limited range of view (the small smartphone screen), and we quickly lose our overall perspective on the task. This makes us think harder (which we all want to avoid doing) and pushes us to make our decision faster.[8, 9] Numerous studies have documented that the small screens of mobile phones create a serious obstacle to users' navigation activities and perceptions.[10] Users need to scroll up/down and left/right continuously within a Web page or an app, which makes it difficult to find target information.[11] Again, the more we need to thumb and tap, the harder we must think.[12] It is also harder for us to remember the content and context of what we have already viewed, which increases the hassle, the burden, and the potential for error.[13] For mobile advertisers, this puts an even bigger premium on saliency.

How to become more salient

Now that we have established that ranking and saliency are even more critical in the mobile ecosystem than on the desktop Internet, let's think about what each position in the ranking hierarchy is worth. This rather academic-sounding question has a strong counterpart in business: How should a firm use insights into ranking and choices to develop an advertising strategy for smartphones? How can brands improve and optimize their investments in search rankings or in real-time customer discounts in order to generate more revenue? How does ranking relate to distance? How can a business become more salient?

My colleagues and I have conducted several studies to answer these questions.

How Ranking, Distance, and Discounts Work Together

Recall the study that underpinned some of the findings about distance vs. discount that were discussed in chapter 5. My colleagues and I collaborated with a major telecom provider in Germany to design a large-scale real-world experiment to understand saliency by testing the effectiveness of location-based "pull advertising." The study was conducted in 374 cities and towns in Germany and involved the collaboration of more than 3,500 merchants. Merchants uploaded their coupons onto the platform, setting their own parameters, such as coupon expiration dates, coupon discount levels, and usage limits. Consumers were then able to pull these deals out automatically via the mobile app. The app was offered on both the iOS and Android platforms. The merchants could offer two kinds of coupons, one with messages only (advertising) and one with a monetary incentive (a discount).

We tested the effectiveness of these mobile coupons in two ways. We looked at the distance between the customer and the business offering the coupon, relative to other stores. We also looked at the ranking of that coupon within the app, relative to all other coupons. GPS location allowed us to calculate the distance between the potential customer and the physical (offline) point of sale in real time. To test the effect of ranking for the purposes of the study, the coupons were sorted either by distance from the user (in real time) or randomly, depending on which test group the user belonged to. The participants were randomly assigned to one of four groups for the duration of the study (see figure 7.1):

- group 1: coupons sorted by distance, with distance information
- group 2: randomly sorted coupons, with distance information

Figure 7.1
Impact of location-based in-app coupons for 373 cities and towns and 3,500 participating firms. These screen shots show what the participants in groups 1 and 2 would have seen in the experiment discussed in the text.

- group 3: coupons sorted by distance, without distance information
- group 4: randomly sorted coupons, without distance information

New considerations: Money talks

A discount of 10–12 percent is the equivalent of improving ranking by one level. Discounts improve saliency, and now retailers and marketers have a rule of thumb to help them calibrate how much to offer to offset the fact that their coupon or ad appears further down in a search result. Retailers or stores who partner with mobile coupon app providers can leverage the quantified odds ratios to calibrate and improve their conversion chances on the basis of the real-time distance between the consumer and the store, the displayed rank of the offer on the screen, and the actual discount. Let's say that the store positioned fourth and the store positioned sixth on the app's display are close competitors. The store with the lower (worse) position or saliency can offset or neutralize that disadvantage by increasing the relative discount in its coupon, but without overdoing it.

The good news is that the study shows that when we are actively searching we are eagerly clicking on the mobile offers that result from our searches. Click-through rates are much higher for mobile pull ads than for mobile push ads, though the rates varied considerably across the four groups in the study.[14]

Two surprising findings

The study had two surprising outcomes. The first involves information about distance. When a coupon did not include information about distance, the random sorting resulted in a lower click-through rate (CTR) than the ranking by actual distance. In other words, users were able to detect the random order and found it to be less relevant. This implies that consumers often have a general sense of how far away sellers are if they are in a city they know well. In fact, more than 90 percent of all the participants in the experiment used the app within a relatively tight range of 10 miles.

The second surprise involves how the number of impressions per session affected consumers' behavior. The number of impressions per session had a significantly negative effect on the probability of redeeming coupons. The number of previous clicks also has a significantly negative effect on consumers' coupon choice probability.

Why does the number of impressions negatively impact the probability of redeeming coupons? Searching wears us out. This seemingly simple finding gives rise to many questions with financial implications for businesses. First, it raises the legitimate concern that too many advertising exposures can backfire. Advertisers make their target customers numb, then immune. Even when pull advertising creates a more favorable environment, business will need to optimize the number of ad exposures. This raises an obvious question: What is the optimal number of ad exposures for a given consumer for a given brand? When is an extra impression one too many? My own work (as described in chapter 6) has demonstrated that number to be between 2 and 4, depending on which stage of the purchase funnel the consumer is at (fewer is better the earlier the stage).[15]

Knowledge of the interplay of all the factors discussed here in part II will help with that optimization challenge. In addition, there is a creative challenge to explore further. If the coupons consumers see do not make a good first impression, they become less likely to click on that advertisement or coupon again. First impressions are critical in this attention-deficit economy.

The experiment mentioned above had a broad and deep enough set of data to yield statistically significant findings for many product and service categories, including multi-media, health and wellness, apparel, groceries, beauty, café, restaurants, clubs and bars, fashion and accessories, and leisure and sport. These findings led to practical insights that managers in these industries can use as an orientation point for rethinking and refocusing their advertising strategies. The findings also provide clear and explicit guidance on the breadth of levers available to optimize a mobile advertising campaign, and how those levers work both in isolation and in combination. These findings create opportunities for well-managed dynamic pricing strategies based on the combination of display rank and real-time distance between a user and the store. The face value of the discount becomes your key tactical weapon to neutralize any ranking disadvantage. There is plenty more to explore and learn, but these data-driven insights can complement gut feeling and intuition with clarity and an extremely high degree of confidence.

It would be unwise to generalize the results of this study to each and every business context. Nonetheless, the findings and insights may help to inform and inspire managers to think about the economic potential of becoming more salient.

Other business implications

The topmost slots on the screen of a smartphone have greater value to retailers, and will continue to attract premium prices. I can imagine the emergence of a new business model for monetizing the space within an app that involves app providers auctioning off the specific ranking slot, in the same way it is done for search engine advertising campaigns. This would result in a higher click-through or impression-based price per ad.

I expect that saliency-based pull advertising providers will implement auction mechanisms to allocate the top-ranked promotions optimally within "distance buckets"—for example, 0–100 meters, 100–500 meters, 500 meters–1 kilometer, and so on. Bid prices would be driven by previous ad performance in a given geographic area, as measured by impressions or clicks (akin to the Quality Score algorithm deployed by Google) and also by actual usage frequencies, which can drive up the value of area-specific and ranking slots.

Saliency is also playing an increasingly crucial role in discovery on mobile app stores. The two major app stores offer more than 3 million mobile apps, and 63 percent of apps get discovered through app searches in the stores. Breaking through is one of the biggest challenges facing mobile app publishers today.[16] The higher an app ranks in an app store's search results, the more visible it is to potential customers. This greater visibility leads to more traffic. Hence it is in their best interest to strive for more saliency. A related development toward increasing saliency is app indexing, which is the process of getting an app into search engine's index in order to easy discovery. As consumers engage more and more with mobile apps, I can imagine many different kinds of opportunities that will emerge for apps to dominate websites in the search engine results page on a mobile screen.

Food for Thought

Saliency is about standing out in a world full of impatient, in-a-hurry consumers searching for answers on a device. This can make it harder, not easier, to find the "best right answer." This chapter offers baselines to build on, but saliency-based mobile advertising in all its forms is still in its infancy. We have a long way to go before mobile advertising—and our knowledge of its effects and potential—reaches maturity.

In the case of smartphones, limited screen size is an important factor. Depending on the model, the screen area of iPhones and Samsung Galaxy devices can differ by as much as 50 percent, and the screen of a personal computer is many times the size of either phone's screen. This significantly reduces the number of offers viewable on the screen, which makes customizing or targeting an advertisement on a smartphone difficult. The ad's position on the store produces an interesting set of three-way interactions for marketers to exploit. Today, advances in neuroscience and eye-tracking technology are looking at what attracts our attention as consumers and what leads to action. These will help researchers optimize Web design for both large and small screens. In the meantime, as brands jostle for attention with their mobile coupons, the topmost slots on the small mobile screens have greater value to retailers and can be priced at a premium.

The world of location-based advertising has changed beyond recognition since the times when travelers' advice for local food and lodging recommendations relied on the knowledge, bias, and whims of a desk clerk on the side of an interstate highway. Fundamental human behavior may not have changed to the same degree, but our ability to use data, to explore data, to understand data, and to influence consumers' thinking has never been greater, nor have the financial implications and the need to manage them.

Takeaways for Firms

- Start using the baselines in this chapter to test, and learn, and gain greater competitive advantage by exploiting the non-trivial effect of the saliency of your offer on a mobile platform.
- Work on improving the saliency of your offer on a mobile platform, because local searches on mobile devices are more likely to lead to imminent in-store traffic and sales than those on desktops.
- When working with an aggregator of multiple, often competing, offers, track the effect of your rank on mobile ad impressions and click-through rates, and correlate these with actual in-store redemption rates.
- When jostling for saliency in a competitive context, use different discount levels depending on the ranking (saliency) of your offer on the mobile screen, and quantify the resulting coupon redemption rates. A lower rank on the screen can be compensated by a higher discount. Increasing distance between a business and a consumer can be compensated by a higher discount.
- Compare consumers' response rates across devices, because the same exact ad or the same exact coupon will have very different response rates when displayed on a smartphone than on a desktop, based on how salient the offer is on each device. Explore the presentation of ads using eye-tracking technology and use the latest advances in neuroscience.

8 Crowdedness: Why Scarcity of Space Matters

When I think of crowdedness in any given context, it immediately conjures a lot of negative perceptions in my mind. I imagine people in airports, train stations, shopping malls, stadiums, and subway trains jostling, elbow to elbow, for an extra inch of space. But might there be a silver lining in this dark crowd? Can firms use crowdedness to create a new kind of force? If so, what might be the key ingredient in this crowdedness force?

Well, the same ingredient that has created a paradigm shift in marketing in the 21st century will also influence how both consumers and marketers leverage information on crowdedness in the smartphone era and beyond. That ingredient is the availability of data on temporal consumer agglomeration patterns that can now be available in unprecedented amounts and depth from public sources. This trend shows no signs of subsiding, if recent actions by the government of Singapore are any indication.

As part of its Smart Nation program, the government of Singapore is "deploying an undetermined number of sensors and cameras across the island city-state that will allow the government to monitor everything from the cleanliness of public spaces to the density of crowds and the precise movement of every locally registered vehicle."[1] In Singapore, the laws are so broadly phrased that the government can obtain access to sensitive data such as text messages, e-mail messages, call logs, and Web-surfing history from Internet Service Providers (ISPs) without a court's permission. Granted, Singapore's government enjoys a special relationship with its citizens because Singaporeans don't seem to mind more surveillance (less privacy) in exchange for better security.[2] But this initiative should provide additional evidence that the tradeoff I have spoken of throughout the book—the exchanging of personal data for better products, services, and guidance—can work to the benefit of all parties.

"The opportunities are endless," said one official from IBM in Singapore.[3] Another way to put this is to say that this enormous set of data may uncover occasions which we never suspected before, or which we have theorized about but could never understand or capitalize on. One such occasion is crowdedness of a user's immediate vicinity. A few years ago, my colleagues and I became very curious about the effect of crowdedness on consumers' propensity to respond to mobile offers.

Not all crowds are created equal, of course. There are many crowds that are pleasurable to be a part of. Think of 100,000 people in the stadium at a college football game, 90,000 in Wembley Stadium for a soccer match, or 20,000 in a sold-out arena for a rock concert or a championship basketball game. In other situations we have crowds that many would consider to be necessary evil or an outright nuisance. In fact, if you peruse the sociology and psychology literature on the topic, you will quickly get the impression that crowds are one of the more destructive forces that humans confront regularly. Studies over the last 50 years have linked crowding to physical and mental diseases and juvenile delinquency.[4] Crowding can increase stress,[5] frustration,[6] and hostility.[7] Individuals may consider themselves more anonymous in crowds, which can reduce their social interactions and fuel antisocial behavior.[8]

Conventional Wisdom about Crowdedness

How do people respond when they are in these unpleasant crowds, the ones we don't want to join? Our reaction in shopping malls and similar environments is clear. Crowds have been shown to make us want to escape. In retail settings it can provoke negative attitudes toward the store.[9] Researchers have demonstrated that when consumers perceive a threat to their freedom due to confinement in tight spaces, they may react by engaging in actions designed to reassert their freedom.[10] More generally, we become nervous and anxious, so we try to evade the crowds and reduce our exposure to them by reducing the time we spend shopping.[11] Crowding prompts shoppers to rely more on familiar brands, avoid interaction with store employees and others, and reduce their shopping time.[12] Overall, these studies have largely shown the negative effects of crowdedness. Hence, historically crowdedness has been undesirable, at least for firms trying to sell.

A couple of decades ago, our evasive maneuvers had to be purely physical ones. We walked out of the store, and left the crowd behind, as quickly as possible. As the 21st century unfolds, we have a different escape route when elbow room gets tight. It's right in our hands.

The smartphone gives consumers a way to escape their surroundings. Consumers immerse themselves in their mobiles phones "as a means of escape, a way by which a person can avoid unwanted encounters" and gain a sense of control over his or her space and privacy.[13]

The smartphone serves as our window to the world, but the window works both ways. The smartphone also allows advertisers a way to approach consumers at stressful times as well. If the advertiser knows how crowded a consumer's surroundings are, perhaps a mobile message or advertisement can provide a safety valve or much needed diversion for people? This scenario provides some validation for the second behavioral contradiction from the introduction. We will indeed welcome advertisements if they are helpful, even if our usual response is to find advertising annoying.

To test this idea in a valid and meaningful way, we needed to find a place where that annoying, stressful kind of crowd occurs frequently and also universally, meaning the kind of crowdedness that is a common experience for many consumers worldwide, regardless of country.

The answer to this question was right beneath our feet: *the subway.*

If you have ever ridden a rapid transit system during rush hour, you know exactly what this crowdedness is like. In most American cities, people spend a considerable amount of time commuting, averaging 48 minutes each way, according to Census Bureau reports.[14] When we commute on a subway, our field of vision takes in a few hundred people, counting those on the platform and in our car. Have you ever thought about how many people ride a subway per day? The numbers are staggering.

In 2015, an average of 5.65 million people rode a subway in New York City *every weekday.*[15] Total ridership for 2015 was 1.763 billion. As large as that volume is, it placed New York in just seventh place worldwide, behind Tokyo, Beijing, Seoul, Shanghai, Moscow, and Guangzhou. Rounding out the top ten behind New York were Mexico City, Hong Kong, and Paris.[16] Even the famous London Underground did not make the list. These top ten lines alone have an annual ridership of 22.7 billion people, which is more than three times the world's population.[17]

As more cities facilitate the use of mobile devices in subways,[18] it is clearly an understatement to say that a daily commute creates a unique marketing environment. Since the average public-transit commute time for Americans is 48 minutes each way, it is no wonder that the co-author of that study said that "considerable amount of down time during daily commutes could be a gold mine for marketers."[19]

But is it really the case? Is marketing via mobile advertising really *more* effective simply because we are crammed together next to strangers in public transit for between half an hour and an hour every day?

Impact of Crowdedness on Mobile Purchases

In a study we published in 2016, Michelle Andrews, Xueming Luo, Zheng Fang, and I cooperated with one of the world's largest telecom providers to test this idea in a major subway system. The underground subway in that city (no, it wasn't Singapore!) is equipped with mobile signal receivers and allows a passenger to use his or her mobile device throughout the course of a commute.[20]

Our goal was to test how crowdedness affects a consumer's responsiveness to mobile advertisements. We defined crowdedness as the number of people per square meter. If you don't ride a subway but you have tiled floor nearby, you can get a sense for this by standing at the center of a 3-by-3 section of typical vinyl or linoleum tiles and imagining yourself surrounded by more and more people.

The Wi-Fi sensors in the city's subway system would enable passengers to have Internet access and allow us to test their responsiveness to mobile offers as the number of people per square meter increased. In theory, there may be an upper bound. If you invite a half dozen of your colleagues to stand on that 3-by-3 section of tile with you, you may not be able to retrieve your phone from your pocket, let alone use it for any advanced function. In Tokyo, the cramming can reach eleven passengers per square meter. In Hong Kong, authorities are considering removing seats from subway trains to give commuters "more room to interact with their devices."[21] In the city of our study, however, the highest level of crowding we observe is five passengers per square meter, so that would be the upper limit.

The cooperation of a telecom provider in Asia (by far the largest one in the country) was essential for the study, not only because we needed to target randomly selected consumers throughout the day but also because we needed to know exactly when they were in the subways. Cellular technology allowed us to record the numbers of mobile users in real time, and with the exact dimensions of the trains, we could determine people's spatial proximity to one another to quantify the level of crowdedness.

For each day of our study, we sent advertisements for various services at five different times, each corresponding roughly to a typical phase in a day's cycle. Within each time cycle, the provider sent text messages to two or three trains, for a total of 14 trains a day. Because the average subway commute is 30 minutes, the reply time was restricted to 20 minutes to ensure that most commuters would respond to the text message while still in transit. The train schedule was:

- 7:30–8:30 a.m. (morning rush hour)
- 10 a.m.–noon (morning lull)
- 2–4 p.m. (afternoon lull)
- 5:30–6:30 p.m. (evening rush hour)
- 9–10 p.m. (after-dinner traffic)

The targeted population of over 15,000 users had not previously subscribed to any of these additional services, nor had they received a similar text message from the wireless company. This helped us rule out potential carryover effects of earlier marketing campaigns.

New considerations: Finding value in crowdedness

The results suggest that, counter-intuitively, commuters on crowded subway trains are about *twice as likely* to respond to a mobile offer by making a purchase as commuters on non-crowded trains. The response rate across the entire sample was 3.22 percent.[22] That rate may seem low at first glance, but it is fairly high compared with the 0.6 percent response rate for mobile coupons and the 1.65 percent response rate for the effectiveness of location-based mobile coupons.[23]

On average, across all the trains, the purchase rates measured 2.7 percent with two people per square meter, and increased to 4.1 percent with five people per square meter. These results apply after we controlled for peak and off-peak times, weekdays and weekends, mobile phone usage behaviors, and randomly sending mobile ads to users.

Another fascinating finding is that the responsiveness is not linear, as the crowdedness increases. If we set a baseline for crowdedness at two people per square meter, the likelihood for a mobile purchase increases by 16 percent when the crowdedness doubles to four people per square meter. But when the crowdedness goes from four people per square meter to five, the likelihood jumps by 47 percent. The more crowded the train, the greater the increase, though because of the travel patterns on this system we were unable to test a limit beyond five people per square meter.[24]

Serendipity granted us a few more surprises during the experiment. On one day, roadblocks due to a government motorcade (a high-security police escort for a politician) closed down streets and caused an unexpected spike in ridership for some periods in the subways. In another case, a train experienced a delay because a passenger's backpack got stuck in the door of a train, resulting in changes both in the distribution of passengers and the length of time they spent on the trains. In those situations we also offered a different but similarly priced product. This enabled us to identify whether purchase variation stemmed from passenger heterogeneity or variation in the level of crowdedness. In all cases, the general direction of the findings held true. Greater crowdedness, whether anticipated or unexpected, significantly increases the likelihood that someone will respond to a mobile advertisement.

Explaining this behavior

Because crowding is often associated with negative emotions such as anxiety and risk-avoidance, the findings reveal a positive aspect of crowding: Mobile ads can be a welcome relief in a crowded subway environment. The next challenge is to explain what causes this higher responsive rate. With its customer service call-center, the wireless provider could survey passengers. So we subsequently surveyed some of those passengers who purchased the promotion. The surveys can help identify possible reasons that crowdedness affects purchase likelihood, after matching the purchase records and measured crowdedness from the field data.

The answer appears to be two words: *mobile immersion.*

As more commuters occupy the train, individuals' personal and physical spaces are invaded. This spatial limitation poses a behavioral constraint that leads to a reduction in the outside options of things to do. Behavioral constraint theory[25] suggests that as more and more people invade one's

Figure 8.1
Mobile immersion is our escape in uncomfortable crowds. (Canstockphoto)

personal physical space, people adaptively turn inwards to filter out inputs from social and physical surroundings.[26] To cope psychologically with this spatial loss and avoid accidentally staring at each other, people escape into their personal mobile phone space.[27] As our experiment showed, they also become more susceptible and more welcoming to mobile ads. They spend more time browsing more intensively on their smartphones in an effort to retreat to a comfort zone and avoid eye contact or other interaction with the strangers surrounding them.[28] Turning inward via the smartphone allows us to tune out the outside noise and ignore or suppress our discomfort. This in turn amplifies our alertness toward what appears on our screens, especially if the advertisement is well curated and fits a need.

I should point out that in our setting, subways were not packed to capacity (such as some commutes in Tokyo, with as many as eleven passengers per square meter). In that case, it is logical to speculate an upper boundary effect of crowdedness: Too dense a crowd would negatively affect mobile purchases because congestion would restrict the ability to use a mobile phone.

In essence, the combination of smartphone and crowdedness creates an entirely new buying occasion. This occasion, as I pointed out with the data on subway rides, is a common and frequent experience worldwide. Even if we conservatively estimate a commute or travel time of 30 minutes and assume that half of all rides are taken alone, the top ten subways systems provide marketers with 5.68 billion hours of consumer downtime every year to tap into.

Google has started to provide free Wi-Fi service at several railway stations in India. As of August 2016, about 2 million people were logging on to the free high-speed wireless access deployed by Google and RailTel at 23 major railway stations across India, and there were plans to expand to 100 stations by December.[29] Observers point out that this free Internet is much faster than the 3G Internet that is most widely used in India. For businesses and brands based in India, this could be a neat opportunity to tap into the Wi-Fi-based system in order to leverage the "crowdedness" force and send relevant messages to passengers. Because these trains are over the ground and the routes and times are fixed before hand, this kind of targeting could also be combined with other forces such as "location," "time," and "weather" to create a much more powerful and appealing offer.

New considerations: Distinguishing between commuters and non-commuters in the crowds

In any crowded situation, there are two broad categories of folks: the ones who are commuting from home to office (or vice versa) and those who aren't. Might people in these two groups have different propensities to respond to mobile messages sent by businesses?

Recall that in chapter 5 I described a recent study in which my colleagues and I worked with the largest mobile service platform provider in South Korea and the public transit system authorities in Seoul. Our goal was to send coupons to a variety of commuters in South Korea (the ones taking the trains, buses, etc.). Among our everyday activities, commuting is a source of stress and is associated with varying levels of frustration, annoyance, depression, and anger.[30] Researchers have reported that stress during commuting is induced by varying environmental factors such as noise, crowding, and air pollution. On the basis of these various psychological and physiological theories, we hypothesized that the stress of commuting increases the frequency of the redemption of mobile coupons.

Two national retail chains provided several food and beverage products for promotional purposes. A total of 71,828 coupons were distributed to 44,622 users randomly selected out of a sample of 1.27 million users. The coupons were distributed within a public transportation mobile app. The most important feature of this app is a public transit guidance function that provides users with several fast route options and estimated arrival times from their departure point to their destination point on the basis of real-time traffic conditions. Upon using this app, users are requested to register their home and work addresses. When using the guidance function on commuting routes, users can simply tap "guide me to workplace" or "guide me home." The application then guides the users from their homes to offices, or vice versa. By contrast, when users wish to be guided on routes other than commuting, they must add a *temporal route* by registering a point of departure and destination. They can then activate the guidance function on the temporal route by tapping "guide me on the route." This feature enables the app to track their regular commuting routes vs. the days they are deviating to a temporal non-commuting route. The application also accurately identifies the location of the user (i.e., bus, subway, or street), and then alerts the user when and where to board, exit, and transfer.

Using a carefully designed randomized field experiment, we randomly selected subjects who used the guidance function of the application, either on commuting routes (the treatment group) or non-commuting routes (the control group), and sent them mobile coupons. When a user activates the *guidance function* on commuting or non-commuting routes, a pop-up window with the message "check the gift" is opened randomly, and the mobile coupon is downloaded to the smartphone of the user. A mobile coupon is an image file that contains the picture, name, expiration date, and unique bar code of a product. Using this bar code, users can exchange their coupon for the product, at no extra charge, at any selected store before its expiration. We can also identify whether, when, and where users have exchanged their mobile coupons by using these bar codes.

We found that coupons sent to commuters had a significantly higher probability of being redeemed than the same coupons sent to non-commuters. Marketers can increase the offer response rate by focusing on specific periods in a given day when commuting stress is relatively high, such as rush hours. The average redemption rate for commuters was 31.6 percent, which was 1.76 times that of the non-commuters.

We found that there are two different ways to induce people to respond to mobile offers. First, firms can vary the number of coupons sent to a user. Second, they can increase the validity period for each coupon.

We also found that commuters respond extremely well to a single coupon and negatively to a promotion with more than one coupon. They show signs of efficiency and focus. Non-commuters, on the other hand, respond extremely well to multiple coupons because they benefit from variety and choices.

And we found that commuters respond extremely well to offers with shorter redemption times. For example, offering discount coupons for which a day or a few hours remain before expiration is an effective strategy for commuters. Non-commuters, on the other hand, respond extremely well to offers with longer redemption times. Coupons with relatively longer expiration dates (i.e., one week) work very well on this sample.

Firms can enhance the effectiveness of their mobile marketing strategies by exploiting commuting as a target environment.

Food for Thought

Much of the literature on crowdedness emphasizes its negative effects, but this new research shows that targeting consumers in crowded contexts with mobile promotions may serve as a welcome relief. The same mechanisms appear to apply for people on a crowded bus, a crowded road, a crowded train station, or a crowded airport. Even when the consumer is outside the confines of public transportation but in a crowded venue full of strangers, such as Times Square in New York, they may respond better to coupons or offers from nearby stores because they offer a physical refuge.

Previous studies by my colleagues and I have shown that consumers are more likely to click on links to stores close to them and on higher-ranked links on their mobile screens. Other studies supported this by showing that either purchase likelihood or mobile coupon redemption rates increased, the closer consumers were to a store and the higher the offer was displayed on the screen, conditional on the discount.[31] If a nearby store can identify a potential customer in a crowd, their success rate may be higher, all things considered, than by targeting an equidistant customer who is not in a crowd, or who is in a different crowd. Again, not all crowds are created equal, and anonymity matters. People in a crowded restaurant or stadium may focus on food, companions, or the game in progress, rather than the ads. Moreover, people may cope with crowdedness differently depending on their preferences and living circumstances.[32]

It is important to distinguish between commuting peak hours and non-commuting peak hours, and design mobile marketing strategies accordingly. Firms need to plan the optimal number of coupons on the basis of whether they are targeting commuters or non-commuters. One coupon for commuters and up to five coupons for non-commuters is optimal. In addition, mobile promotions with shorter redemption periods are more effective for commuters, and those with longer redemption periods are more effective for non-commuters.

I can imagine many new entrepreneurship opportunities will emerge in order to facilitate established firms to be able to comprehensively tap into this force in different contexts. Many established companies in the consumer packaged goods industry, in retail, in fashion, in apparel, in fast food and beverages, in travel and hospitality, and in other industries will have to collaborate closely with new vendors to will have build this infrastructure, not only to be able to predict and estimate crowdedness in any given location but also to be able to deliver the ads in those places. What we have showed in these two studies is only the tip of what is likely to be a tremendous opportunity for monetization.

Because there are these micro-moments in the crowdedness space, it makes sense for businesses to make these finer distinctions between commuters and non-commuters.

Takeaways for Firms

The stakes are high. One estimate claims that mobile ad spending will reach as much as $100 billion by 2018.[33] Given those projections, firms have a vested interest in reaching consumers when and where they are most receptive to mobile ads. To get a bigger bang for their buck, firms should leverage the immediate crowdedness of a consumer's vicinity. Crowdedness is a powerful force that they can ignore only at their own peril. Three takeaways should be noted:

• As people stand elbow to elbow in many crowded contexts, a mobile ad has been proved to be a welcome diversion. Redemption rates on mobile offers sent in crowded settings can be twice as high as those in non-crowded settings.

- When examining the effectiveness of mobile offers, it is important to distinguish between commuting hours and non-commuting hours, and between commuters and non-commuters.
- Depending on the target group, firms can increase redemption rates either by varying the number of coupons sent to a user or by changing the redemption period for each coupon or by doing both simultaneously. Commuters respond best to single coupons with shorter redemption periods. Non-commuters respond best to multiple coupons with longer redemption periods.

9 Trajectory: Watch Where You're Walking

Tom and Rachel live in Nashville, Tennessee. Tom is in his late thirties and works as a marketing executive for a small music label. Rachel, likewise in her late thirties, works for a bank downtown. They are both currently standing on a corner at the intersection of Grand Avenue and 21st Avenue South on an unusually warm spring day at 5:45 p.m. It is coming up on dinner time, and they dine out frequently between Music Row and Vanderbilt University. If you are a local restaurateur or merchant, it would help to know what their plans are for this evening. Or perhaps—in the absence of such knowledge—you may be able to wield some influence over those plans.

Tom and Rachel are ripe for targeted, customized advertisements from all the nearby restaurants. In fact, any business that can home in on who they are and where they are has a potentially lucrative target in sight.

The ability of location-based advertising (LBA) to target such a desirable audience directly gets us one step closer to the holy grail of advertising, which is to deliver the right advertisement for the right product to the right person at the right location for the right occasion at the right time. It is no surprise that LBA is becoming an integral part of any brand's or retailer's advertising mix.

The future looks very bright. Yahoo's vice president of strategic insights and research proclaimed in 2015 that "the next generation of location-based advertising offers a huge opportunity for brand marketers to provide consumers with relevant and useful ads. Mobile is changing everything, and as marketers, we continue to uncover more and more possibilities."[1] In her view, the key factors are the where (location), the when (time of day), and the who (demographics and purchase history) of the potential target.

For Tom and Rachel, all that information is accessible if they have their smartphones turned on and enabled.

I have already described the location and the time of day. This chapter is about the demographics and the behavioral history. But this history is not just about knowing your age, gender, income levels, and the basket of goods you have been buying in the past. In addition to all that history, it is about *tracing exactly how you went about that process when you came to shop offline.* What if I told you that the future looks very bright, but not for the reasons the Yahoo executive mentioned? If LBA is an important step toward that holy grail of advertising, what if you could make a quantum leap?

There is an approach that will make the future of mobile advertising shine even brighter. It already exists, and it has proved its potential, as I will demonstrate later in this chapter. A major force in mobile advertising in the coming years will not be personal purchase histories, or where someone is now. *It will be in predicting where someone will be on the basis of where he or she has been.* The future of mobile advertising lies in making sense of the rich data in people's pasts, not in relying solely on their present state, in order to predict where they are headed. Brands need to pay careful attention to the words of the hockey player Wayne Gretzky: "I skate to where the puck is going to be, not where it has been."

In the online world, it took us only a few years to get accustomed to, and often even embrace, the idea that firms—including e-commerce firms, search engines, and website publishers—can track our browsing behavior and predict our next steps. A similar revolution is about to hit us offline. The springboard for this revolutionary leap is the individual's *trajectory*. As was noted in the introduction, an individual's trajectory is the physical, behavioral trace of his or her offline movements. Firms can measure when we walk past their physical stores, when we come through the front door, when we walked up to the second floor, and so on.

Comparing location-based data with trajectory-based data is like comparing a snapshot with a high-resolution video clip. A snapshot can capture the present moment in rich detail, but a video clip allows us to take motion over time into account, to see patterns, and to pick up other clues that may be influencing a person's behavior. Over time, the length and resolution of video clips is likely to improve, putting richer and more precise series of data in the hands of people designing marketing, branding, and advertising campaigns. As they combine the insights from the video with the snapshots, they can then begin to paint a better picture of consumers that will beget a richer shopping experience. This is what makes it possible to create the equivalent of the Marauder's Map in the Harry Potter books, but improve on it by giving it a large memory cache.

Let's go back to Tom and Rachel on their nice early evening in Nashville. Imagine if you also knew that their movements that day have not only been virtually identical but also have been very slow. It is obvious that they are not driving. Even at a walking pace, though, they have made frequent stops and starts. They clearly aren't following a steady, smooth pace. If a marketer had been able to trace their movements over a longer period and view their individual trajectories, two things would have become apparent: First, they are walking the dog together, as a couple. Second, on this particular trip, as is their habit, they will buy coffee at Starbucks, but will buy no food or snacks. Because of the unseasonably warm temperatures, the chances are good—though not 100 percent—that one of them will order an iced coffee instead of a hot one. Then Tom and Rachel head home. They do not go anywhere for dinner yet.

The rich trail of data on trajectories is valuable because of the patterns and tendencies it reveals. It brings both intuitively clear tendencies and trickier subtle ones to the surface. These patterns reinforce one of the four contradictions mentioned in the introduction: People seek spontaneity; however, they are far more predictable than they think they are, and they value certainty. Our personal trajectories are littered with clues which reveal that our perceived spontaneity is often little more than a different flavor of the same predictable patterns. We have consistent habits that apply even when we enter new situations.

To figure out whether trajectory-based advertising works in real life, we need to define more precisely what we mean by trajectory so that we can measure it consistently. Then we need to conduct some experiments. And that is exactly what we did.

Making "Trajectory" Measurable

Trajectory has three dimensions: time, route, and velocity. Time includes the starting and ending point of the trajectory and the day of the week. Route is not location itself, but rather a way to determine how similar someone's spatial trajectory is to others'.

Imagine two individuals, José and Tara, in New York City. If we find that José and Tara have both taken the same subway train to Penn Station on a Monday evening, that doesn't necessarily indicate an overlap in their trajectories, even though they occupy more or less the same space and got there using the same route. Hundreds of New Yorkers, if not thousands, followed the same route that day, so we have to adjust our analyses to take the crowdedness and the popularity of a given route into account.

But if José and Tara then go upstairs to Madison Square Garden to watch the Knicks game, then both dine at the same small restaurant, their trajectories have shown a meaningful similarity and overlap. That doesn't necessarily indicate that they are friends or even acquaintances, though we may potentially deduce that from where they sat at the game, but they appear to have some behaviors in common and to be in the same customer segment.

Finally, velocity contains information about how fast the individuals are moving. The information buried in these patterns is vital. Among other things, it helps us understand whether someone is a focused shopper who is on a mission for one particular item, or whether someone is an explorer, waiting for something to catch his or her eye.

Underlying these three dimensions is a fourth and far more granular dimension called *semantics*. Semantics takes a number of factors into account, such as the likelihood that someone may visit a certain store, how much time they spend there, how much time they spend moving to another location, and how related or unrelated those two stores are. Going from Dick's Sporting Goods to Academy Sports is a trip within a related product category. Going from Dick's to Staples isn't.

Figure 9.1 is based on fast and accurate monitoring of group walking patterns in a dense urban space.[2] It shows some sample trajectories of members in a group and shows how we can compare individuals' movements. The upper left graph is based on smartphone accelerometer streams and shows how one can extract motion features that indicate stationary versus motion states for an individual. For example, owing to the design of the mall, people walking in a group often take certain turns together. As the upper right graph shows, this is captured by turn similarity obtained from the compass degrees of a smartphone. The graph at the bottom of the figure captures how members tend to move between different floor levels of a mall in a

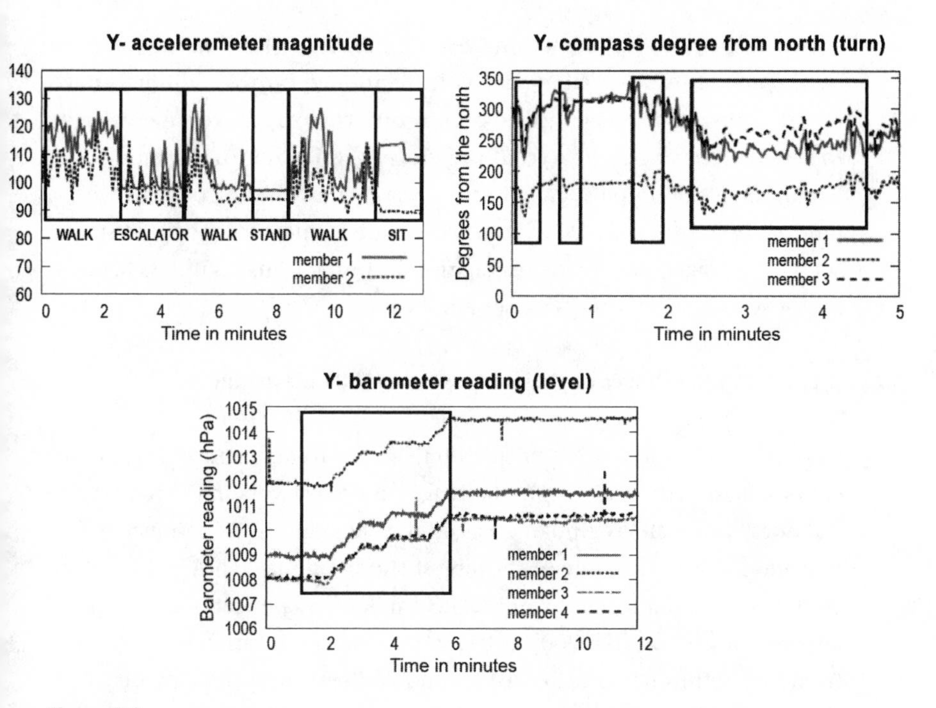

Figure 9.1
Ways we can make trajectories visible and comparable using data from a smartphone's accelerometer, barometer, and compass. (adapted from GruMon by Rijurekha Sen et al.)

coordinated way. In a multi-level building, these coordinated level transitions can be detected using smartphone barometers.

Focus and exploration are two shopping "stages." Several marketing and psychology studies over the last 100 years reinforce the same theory: that we progress through different stages in our thought processes, and that the progression leads to our decision on whether to buy something.[3] This theory has its basis in the way we consumers process information as we make decisions.[4]

When we are in an early shopping stage, such as "exploration" or "awareness," we are more likely to buy something on impulse.[5] If we are exposed to something randomly during this phase, it may trigger an unplanned purchase because we think of a new need or recall one we had temporarily forgotten. I'm sure that almost everyone has had an experience in which something he or she saw or heard indirectly provided a reminder or some encouragement to buy something, even if the connection was indirect. For example, you hear the Chevrolet Volt mentioned on the radio in a news report during your homeward commute, and you may make a mental note to buy batteries when you go to the grocery store later that evening. The likelihood for an impulse purchase is even greater when a consumer sees a random item on sale.[6] However, when we are in the later shopping stages, such as "engagement" or "consideration," our focus makes us less likely to response to random things we see and hear.

Uncovering the Power of Trajectories for Mobile Marketing

On paper, the idea of using people's trajectories to help customize and target mobile advertisements is tantalizing. But does it work in the real world? If it does, how well? To find out, Beibei Li, Siyuan Liu, and I conducted a set of elaborate studies at one of the largest shopping malls in Asia.[7] The mall contains over 300 stores spanning 1.3 million square feet. On average, it attracts more than 100,000 visitors per day. The experiment took place from June 9 through June 22, 2014, and 252 stores in the mall participated. We then analyzed the data from 83,370 unique user responses, using data from indoor positioning systems such as Wi-Fi systems. (Outdoor positioning systems such as GPS or cellular towers would not have worked so well indoors, and as a result the location data wouldn't have been precise.)

The design helped to ensure that we would have a rich and diverse set of data to work with, not just a large one. In many large malls, a broadband Internet connection such as LTE or 4G does not work very well, especially when we go deep inside. This is important because (as was mentioned in chapter 5) many customers like to engage in mobile showrooming.[8] In order to deal with this inconvenience, many malls and large department stores have begun to offer free Wi-Fi access to their customers.

In 2013, the *New York Times* ran a story about how the retail chain Nordstrom uses Wi-Fi to track shoppers' movements around a store. Nordstrom installed sensors in its stores to scan for smartphones. Any phone that had its Wi-Fi turned on would get picked up by the sensors, which would then make note of the device's MAC address (an address that's unique to a phone) and use it to identify and follow the device as it moved about the store.[9] Mass-market chains such as Family Dollar and specialty retailers such as Cabela's, Mothercare, Benetton, and Warby Parker have been testing various kinds of indoor positioning technologies, typically based on widely available wireless radio technologies (such as Wi-Fi and Bluetooth) and on short-range proximity sensor technologies (such as RFID). They are using the data to design store layouts and to offer customized coupons.

At the entrance of the mall we studied, customers were offered the option of accessing free Wi-Fi service. At the same time, they were notified that logging on to the Wi-Fi would enable the mall to monitor their shopping trajectories, and that in return for sharing their data they would receive personalized coupons and ads as they went about their shopping. Full transparency between consumers and brands with respect to the use of their data was critical for us in order to evaluate the extent to which consumers were willing to share their personal information and be monitored in real time in order to receive relevant offers on mobile devices.

My initial expectation was that a very small number of customers would opt in to this kind of explicit data-sharing relationship with the mall. But as it turned out, more than three fourths of customers basically told us "Take my data and give me an offer I can't refuse." This brings me back to the fourth behavioral contradiction I have emphasized throughout this book: People think they care a lot about data privacy, but they are willing to use their data as currency. This increasing recognition of give-and-take between customers and businesses is a good thing. If consumers want to avoid intrusive or irrelevant ads, they should help businesses help them. By sharing their data, they make it much easier for businesses to curate relevant offers for them in a way that does not make the whole process annoying or overwhelming. To businesses, I repeat: *Mobile should be used to perform as a butler or a concierge, not as a stalker.*

Once a consumer in our study had connected to the mall's Wi-Fi, we were able to track the detailed mobile trajectory information during his or her visit in the shopping mall with precise time stamps. But before a consumer received Wi-Fi access, he or she was required to complete a form that asked about age, gender, income range, type of credit card (gold, platinum, gift card, other), and phone type (iPhone, Android, other). At each store, when consumers purchased a product, they were required to complete another form, which asked for some similar personal data as well as the amount spent and whether the purchase was related to a mobile coupon. Later we cross-validated the information on the two forms to make sure the information was accurate.

Every day for two weeks we randomly assigned around 6,000 mall visitors to one of four groups:

- a control group, which would not receive any mobile ads
- a random advertising group, which would receive a mobile ad from a randomly selected store
- a location-based ("where you are") advertising group, which would receive a mobile ad based on current *location* information
- a trajectory-based ("where you've been") advertising group, which would receive a mobile ad based on consumer's *trajectory* information.

Consumers in the last three groups received their mobile coupons in text messages linked to their phone numbers, so that the coupons could not be exchanged with others. To reduce the risk of bias, we randomized the 252 stores that took part, representing the range of categories we would expect in any large shopping mall. We also randomized the design of the coupon (e.g., 10 percent off, 20 percent off, 30 percent off, or 50 percent off) for the same store, and even randomized the format ("price 50 percent off" vs. "buy one get one free").

Using previous studies as a guide, we tried to track several aspects that we thought would help us trace individual trajectories. To do so, we deployed ShopProfiler, a shop-profiling system that crowdsources data solely from sensor readings from mobile devices. Data collection is automatic and runs in the background. For a customer, the data show what stores that customer visited, how long he or she stayed in each store, and how fast he or she was walking. From the stores' viewpoint, the mobile-sensing data reveal information about the shop's inside layout and how many people visit the shop in a particular time period.

After the consumers left the mall, we conducted a short follow-up survey via the mobile phone, asking each consumer whether he or she had redeemed any of the targeted mobile offers that were sent to them or wanted to receive such offers in the future, and what his or her degree of overall satisfaction with the shopping experience was. We also asked for a small amount of additional personal information, such as whether the person had been a first-time visitor to the mall, whether he or she had shopped alone or with others, how much he or she had spent in the store that had sent the advertisement, and how much he or she had spent in the mall that day.

To understand the shopper's stage in his or her purchase process, we used a simple rubric: We labeled a consumer who had consecutively visited multiple stores in the same product category "focused." We labeled a consumer whose last few store visits were from different product categories an "explorer." We took this snapshot at least 10 minutes after a shopper entered the mall, but before sending the shopper a coupon. Of course, a shopper could also be "focused" with a to-do list that included many different shopping needs (e.g., birthday card, kid's shoes, shirt for work, wedding gift). These labels were meant for illustrative purposes only, to help us distinguish between these two kinds of customers. We could just as easily have come up with other labels.

New Consideration: The Value of Offline Data in an Online World

If data-driven location-based advertising gets us one step closer to advertising's holy grail, then trajectory-based advertising does indeed bring us one step closer. We learned that trajectory-based advertising—in general—significantly increases mobile coupon redemption rates, but we also learned in what contexts it has its greatest power, and in what contexts it actually underperforms other forms of advertisements. The most compelling insight is that trajectory-based coupons have much higher redemption rates than location-based and random ones: On average, the redemption rate was 35 percent higher than that of location-based ads and 94 percent higher than that of random ads.

Although we found the mobile trajectory-based advertising more effective for attracting the focused shoppers than the other advertising approaches, we also found that random ads and location-based ads were more effective in attracting the explorers. If a shopper is exploring rather than focusing (meaning that the shopper is prone to an impulse or unplanned purchase), random ads outperform trajectory-based ones, especially on weekends. This is one of the biggest arguments for a mindful, data-driven advertising strategy that makes use of all three forms—random, location-based, and trajectory-based—depending on the consumer's context and needs.

All these results hold up just as robustly when we look at individual customers rather than at groups. We find that, on average, trajectory-based mobile advertising outperforms all the baseline advertising strategies, followed by location-based and then random advertising.

New considerations: Efficiency of trajectory-based advertising

Trajectory-based advertising increases consumers' spending and simultaneously makes them more efficient. A shopper who received a trajectory-based ad from a store spends 38 percent more in that store than a consumer who received a location-based ad, and also spends 26 percent less time in the store. The insight into efficiency is especially important, because it emphasizes the benefit that consumers receive when they exchange data with brands or retailers—the value creation, as I often call it. You might recall the second behavioral contradiction from the introduction as well. Yes, we find advertising annoying, but at the same time we fear missing out and dislike the time wasted in the trial-and-error process of searching for what we need. Trajectory-based advertising tilts the scale even more in favor of advertising that is immediately helpful, not annoyingly intrusive.

A shopper who received a trajectory-based coupon spent just under 10 minutes in the store that sent the ad, and spent, on average, $56.78 at the store. A shopper who received a location-based coupon for the same store spent about 13½ minutes in the store, and spent $41.25, whereas a recipient of a random coupon roamed the store for more than 28 minutes and spent just $23.50 on average.

New considerations: Demographic differences

Age and income also make a difference in response rates. Trajectory-based ads work very well on high-income individuals. There is a preconceived notion that this group normally is not responsive to promotions, but in reality we saw repeatedly that this group is the most receptive to highly targeted offers. The high-income group is not as responsive to random ads or location-based ads, but shows an extremely strong preference for trajectory-based ads. This further supports the idea of efficiency, meaning that high-income people appreciate the kind of curated, targeted advertisement that trajectory-based advertising delivers. Shoppers with lower monthly income are, on average, more active in redeeming mobile ads, again regardless of the type of ad. This finding is reasonable, because we assume that low-income customers are often highly sensitive to price and will find any mobile ads attractive if they are price promotions.

Younger shoppers (aged 20–30) are more responsive than older shoppers (aged 40–50+), regardless of the type of mobile ads. Female shoppers tend to spend more than male shoppers in the stores that sent the coupons, but male shoppers are much more responsive to trajectory-based ads than the other two forms. This result is in line with previous findings that men prefer more guidance while shopping than women do.[10] Our result suggests that well-designed mobile advertising campaigns can serve the purpose of providing better shopping guidance.

Can Trajectory-Based Advertisements Really Alter Our Behavior?

An interesting question is whether shoppers fundamentally change their behavior patterns after receiving a mobile ad. Are we really malleable and trainable? Do we have fundamental, fixed traits, or are we subject to influence? These questions are important because the answers will enable us to better understand how to drive the incremental revenue for stores, and to examine the effects of advertising in the short term vs. the long term.

When we did the analyses described above, we divided all the consumers (and their respective trajectories) into ten groups, each of which showed behaviors similar to other members of that group, but different from anyone else. The challenge to test this idea of influence is to see whether receiving a mobile ad causes someone to diverge momentarily from a behavioral pattern, or whether the behavioral change after the ad is so pronounced that the shopper actually leaves one group and "joins" another.

In the control group, which received no advertisements at all, 1.8 percent of customers naturally diverged from their original behavioral patterns in the mall. Interestingly, by comparing the data before and after, we could ascertain that 13.1 percent of customers in the trajectory group changed from their original segment to a different segment, followed by the location group (11.28 percent) and then the group that received random ads (7.8 percent). Customers from the trajectory group, on average, visited the highest number of new store categories (five) after receiving their advertisement. The shoppers in the location and the random groups visited an average of three new categories, and the control group visited only one new category on average. These findings suggests that the effectiveness of trajectory-based advertising lies not only in the way it can make the shopping experience more efficient but also in its apparent ability to nudge customers toward changing their future shopping patterns, generating even more additional revenues for the stores. In other words, the focal advertising store always benefits from well-designed mobile ads.

All these findings underscore several important aspects about our understanding of our smartphone culture, the marketing opportunities it creates, and the attention it deserves. The first point is how dynamic this market is. Location-based advertising is poised for rapid growth, and companies are about to reallocate and optimize their advertising spend. But in industries such as retail, trajectory-based advertising may soon supersede LBA as the primary form of mobile marketing. This is different from saying LBA will become obsolete. Location-based advertising, as the study in the mall demonstrates clearly, also has a positive impact on generating revenue, increasing customer satisfaction, and making people more comfortable with sharing data when they know that doing so can make their shopping experience more efficient. The impact pales, however, in comparison with the performance of trajectory-based advertising.

Which dimensions matter the most?

In the final test we looked at the various dimensions that define a user's trajectory—time, route, velocity, and semantics to see which of these four dimensions had the strongest ability to predict customer behavior. We noticed a couple of important insights that make the business case for trajectory-based advertising even stronger.

Time is a crucial dimension. This is somewhat intuitive. The longer we can observe the trajectory of consumers, the richer the overall mobility information is. The "fine-grained" information derived from the trajectory becomes more precise and its value becomes more significant. When the trajectory is short, however, this fine-grained information may not be significant enough to make a difference vs. the more conventional and affordable coarse-grained data. The semantic dimension such as the probability of moving from one store to the next with or without an advertisement was also a strong predictor of individual future behavior.

The insights are interesting for what they omit. They suggest that location proximity alone is *not* sufficient to understand and predict consumers' physical behavior. I don't say this to undermine the usefulness of location-based advertising, but instead to emphasize that the inherent advantages of trajectory-based over location-based advertising. By comparing the value of information from the different measures of mobility, we find that location proximity is the *less* valuable, while historical movement information is the more valuable in predicting consumers' physical behavior. The advantage of the fine-grained mobility data over the traditional behavioral data becomes even greater when the user's trajectory is longer.

Other examples

Use cases for this kind of customer trajectory mining from mobile data is spreading across the world and firms are beginning to tap into this golden opportunity. The Mall of America in Minnesota—which attracts 42 million visitors a year, including 3–5 million international visitors from Canada, Latin America, Europe, Japan, Korea, and elsewhere—recently undertook an instantiation of this kind of mobile data mining.[11] Ravi Bapna and a group of five grad students analyzed the non-identifiable customer-level data accruing from Wi-Fi access for insight into foot traffic patterns, dwell times in particular locations, and the effect of events on visitor counts. The students identified clusters of shoppers on the basis of their movements through the premises of the mall and extracted insights into activity at different entrance and exit points, into how stores' promotions affect cross-selling, and even into shoppers' affinities for certain retail brands.

I predict that in a few years more malls in the United States and elsewhere will be pursuing kinds of elementary analyses with Wi-Fi data similar to those that the Mall of America pursued. Eventually I expect to see more and more businesses across different industries adopting the level of sophistication in trajectory-based mobile marketing as the projects we undertook in the shopping mall in Asia.

Another interesting example pertains to stores' using cameras with heat mapping to access shopper movement analytics using software for firms like PRISM. At its core, heat mapping uses video images to visually illustrate the location and density of people. Using this heat mapping technology, Combatant Gentleman, an Irvine based retailer, found out that when men enter their physical stores, they flock to cotton suits. That was in stark contrast to their online customers, who primarily flock towards wool items. So when those cotton suits were moved to the front of the store, Combatant got a huge boost in sales.[12] By combining heat mapping with radio-frequency identification (RFID) tags in apparel, retailers can determine which items are being tried on and abandoned, which shelves are luring customers and which aren't, and so on.

Food for Thought

Think back to Tom and Rachel on that street corner in Nashville. It is their patterns over time that would enable us to understand important finer details—walking the dog, Starbucks as the destination, coffee only—rather than act on location or other factors in isolation. It is this information that tells us that Starbucks has an opportunity, while Chipotle one block away is probably wasting its money at that time.

As with many findings in this book, the greatest power comes from the *combination* of forces and data, not from any one force or source in isolation.

Marketers can refine and sharpen their marketing strategies by combining traditional data sources (social media, payment or loyalty cards, etc.) with tracking and localization data from Wi-Fi, Bluetooth, RFID, apps, and in-store video. This allows them to use transaction data for marketing purposes (traffic patterns, pricing, conversion) and not just for operational purposes (stock replenishment, merchandising).[13] Video plays an important role, because it allows the merchant to observe what kinds of customers enter the store (gender, social context), where they stop, and how long they spend at a display in order to link that information to conversion. When people have their phones turned on, the merchant can help them even more as they exchange data. As one marketing specialist put it, "To me, in-store analytics is not about tracking people, it's about helping retailers respond to customers."[14]

Making our trajectories known is not without risk, but we have to keep in mind that the companies who collect and analyze data on our trajectories have a vested interest in keeping such data out of malicious hands and away from malicious uses. Those companies are well aware of that. The good news is that, in the vast majority of cases, data gathering is conducted in a way that protects users' privacy. No personally identifiable information is collected. Retail analytics firms selling these technologies, such as Euclid and Moxie Retail, will integrate their own market data with the retailer's market data to add more contextual information. Euclid, in fact, was part of the group that launched the Mobile Location Analytics code of conduct, endorsed by Senator Charles Schumer (D–New York). The code of conduct requires companies to have consent if they collect personal information, and to post notifications on how consumers can opt out. The Federal Trade

Commission also endorsed the measure for "recognizing consumers' concerns about invisible tracking in retail spaces and [taking] a positive step forward in developing a self-regulatory code of conduct."[15]

Trajectories bring social benefits as well, and those benefits go beyond marketing and commercial purposes. Trajectory data on nurses and doctors can be used to streamline workflow operations in hospitals and health-care centers with the goal of minimizing patient wait-times and maximizing doctor-patient or nurse-patient interaction times. The application of trajectory analyses extends to the analysis of data on passengers' movements in airports. Researchers from Denmark analyzed data obtained from Copenhagen Airport and containing more than 7 million observations from 25 Bluetooth base stations, covering more than 74,000 users.[16] For example, when managing a large airport, it is useful to determine how passengers move between different regions (the tax-free shopping zone, the security area, the immigration and customs areas, the various boarding gates, the lounges, and so on). Minimizing congestion in intra-airport movement and waiting times at strategic points is yet another application of mobile-phone-enabled trajectory analyses. Law enforcement has long used location pattern analysis and forensic psychology "profiling" to determine who might be a criminal and where he or she might go. The trajectory data of mobile phone users combined with social media data also helps police anticipate trouble spots. Police in the California city of Huntington Beach combined forces with the social media analysis platform Geofeedia to look for patterns in social behavior, such as use of the word "gun" or "fight." These keywords can indicate a potential need for increased patrolling.[17] Since the person may be posting social updates from a particular set of locations, or the locations themselves have a history of such words within the geo-fence perimeter, the police can examine the set of locations instead of the person, thus protecting the individual's privacy and the public interests at the same time.

Takeaways for Firms

New technologies such as video analytics, Wi-Fi analytics, and beacons have emerged to help firms optimize their offline store experience and profits. As our data improve stores' understanding of their traffic flows, they will allow businesses to make well-informed business decisions about how to change in-store layouts to minimize congestion, where to place promotions so that consumers can spot them easily, how to increase sales conversions, and so on. When a business plans to take advantage of trajectory-based advertising, it will have to make the commitment and earn the trust from your customers to observe them for longer periods. But one should avoid drawing blanket conclusions. The same technique that excels on weekdays can backfire on weekends. The same technique that excels on focused shoppers can backfire on variety seekers and explorers. The jackpots are there and they are probably larger than one anticipated, especially when one takes customer satisfaction and customer lifetime value into account. Making this kind of quantum leap comes with a cost, but it is worth the investment in the rich, contextual information that makes predictions of behavior more reliable and more lucrative.

Some important takeaways for firms are as follows:

- The smartphone is the glue between offline and digital channels. Much as data on consumers' Internet browsing revolutionized the ability of firms to target users online, trajectory data from users' mobile devices are going to revolutionize brick-and-mortar firms ability to target consumers offline.
- Trajectory-based mobile advertising significantly increases the mobile offer redemption rates and revenues of the store as well as the overall revenues of the shopping mall. It is especially effective for targeting high-income individuals.
- Trajectory-based mobile advertising is very effective for targeting focused shoppers during weekdays. On weekends, trajectory-based mobile advertising becomes less effective for the "exploring" shoppers, in comparison to random ads and location-based ads.
- Trajectory-based advertising can not only make the immediate shopping experience more efficient, but also nudge customers toward changing their future shopping patterns.

10 Social Dynamics: You Are Who You're With

Food courts at large shopping malls can fill up quickly around noon on a Saturday. Let's say you would like to send an advertisement or a push offer to the 100 or so people who are eating lunch. You plan to offer a discount for a consumer electronics store, and the coupon will expire today when the mall closes. Whom would you target? Or put another way: who do you think is most likely to look at and respond to your ad?

To be more specific: Do you target the three college girls sitting together at a booth? Do you target the young mother and her toddler son? How about the two smiling people (probably in their late twenties) who are alternating between sharing their food and holding hands? What about the two well-dressed middle-aged men who are chatting but don't appear to be acquaintances?

There is strength in numbers, as the saying goes. But is there also value in numbers?

For a retailer or a brand manager, whether social groups matter in influencing purchases is an interesting question. But its practical relevance had not been fully leveraged until the emergence of smartphones and the ability to target someone with a customized advertisement depending on where he or she is at a particular time and who he or she is with. Nowadays, answering this question may be critical for the success of an advertising campaign. When a business wants to send a push advertisement, is it better to target individuals on their own or individuals in a social group? And if we do get a rule of thumb, what are the exceptions to it?

These questions becomes very lucrative ones, not just interesting ones, if the differences between targeting individuals on their own and targeting individuals within social groups are substantial. That would make the answers worth knowing, worth investing in, and worth exploring further.

As was noted in chapter 4, the easiest answers are to say "it depends" and to claim that you need more information. Although "it depends" is a fair statement, it sidesteps the growing empirical evidence that there are indeed significant differences in the likelihood that the people like the ones at the food court will respond to your ad. This chapter offers some important additional guidance. We know already that time and location matter, and in general, the people gathered at the food court on this day form a captive and receptive audience, possibly in a crowded context.

But the social company we keep changes our behavior as well, as recent studies have indicated. This is our social context. Beyond our individual location at any given time, our social context influences how we interact in real life as part of a group of friends, as a couple, or with family members.[1] And these behaviors are fundamentally different than how we behave when we are on our own.[2]

One study found that during the "tween" years (defined in the study as the age range 8–12) children begin to replace their family members' opinions about purchases with their friends' opinions.[3] Another study found that people are likely to spend more money when they are shopping with friends than with family. Specifically, shopping with peers increases our urge to purchase, while shopping with family members *decreases* it.[4] However, this difference is greater when the group (peers or family) is cohesive and when participants are susceptible to social influence. Another study showed that shopping with a friend can be expensive for males but not so much for females. Said simply, the authors assert that, relative to women, men spend significantly more when they shop with a friend versus when they shop alone.[5] All these studies confirm the hypothesis that certain behavioral biases creep into people when they are shopping in groups rather than individually.

Shopping among strangers can even trigger a lemming effect. In some contexts, watching strangers shop can make us more likely to make a purchase, too, and even buy the same kinds of things we see others buy. One study observed the buying behavior of airline passengers, who could use the in-flight entertainment system to shop. The study showed that the average number of purchases per passenger rose by 30 percent for passengers who could see another passenger buying something. If they witnessed two passengers making a purchase, their own odds of purchasing would rise by 58 percent. Even more fascinating, if another other passenger bought food, the observing passengers were 30 percent more likely to buy food than those who didn't see someone buy food. A passenger who watched someone buy alcohol would be 78 percent more likely to buy alcohol.[6]

As we have discussed before, the general density of people in a given place also makes a difference even when we ourselves happen to be there only with a small group of people or on our own. If marketers want to measure the effectiveness of their mobile marketing campaign, it is critical to understand how crowded an individual target's immediate environment is.[7] As was noted in chapter 8, my colleagues and I have found that the more crowded the customer's current location environment is, the more likely the customer will be to respond to a mobile ad.

Such real-time, real-world social group dynamics can sharpen mobile advertisers' abilities even further as they attempt to reach the right targets with the right offers. If they understand consumers' preferences more fully given a specific social context, they can use that knowledge to provide consumers a better digital experience. The fact that family and peer groups have varying levels of influence on individual impulse purchasing suggests different promotion and advertising strategies for different shopping groups.[8]

An interesting field of study called human–computer interaction looks at the important ways computers change the way people behave. Unlike other tools such as a coffee cup or a screwdriver, computers and their popular descendants such as smartphones cooperate with people. They learn from people and with people. And as anyone with a smartphone knows, they can also interrupt us because many people forget to change their notification settings at different points in time during the day. Researchers from Carnegie Mellon University's Human-Computer Interaction Institute have studied how people respond to such smartphone interruptions depending on their social context. They have found that when people are together in groups in real life, they are more likely to pay attention to unexpected interruptions or notifications via their smartphones.[9] Furthermore, the academic literature in psychology shows that the size of the group also matters, and that small groups (i.e., couples) are likely to show qualitative differences in behavior from larger groups (e.g., groups with three people or more.[10] Based on these theories, we expected to see sharp differences between in the behavior of singles, doubles, triples, and neighbors.

The challenge for the advertiser is to get access to this kind of social context information in real-time. If consumers have opted in and allowed merchants to collect data about their current movements, merchants can find them through conventional geo-location data. But businesses still have little idea who a person is with solely on the basis of his or her location data. Getting this information through surveys is difficult, because ideally the information needs to be collected frequently and available in real time and in a scalable manner. They will typically have to harness external and secondary sources of data, but that is certainly feasible in many contexts.

New considerations: Impact of social dynamics on mobile purchases

Beibei Li, Siyuan Liu, and I figured out a way to test the importance of social group dynamics in influencing the effectiveness of mobile ads.[11] In an earlier study (described in the previous chapter on trajectory), we had conducted large real-life experiments which demonstrated that mobile advertisers can significantly improve the effectiveness of mobile ads when they incorporate the mobile trajectory information of individuals.[12] Our follow-up study on social dynamics leveraged not only the full historical information on consumers' digitized offline trajectories across different variables, but also the offline social contexts of consumers in order to infer preferences and find ways to improve mobile advertising. We set up an experiment that allowed us to make inferences about an individual's activities on the basis of patterns in the individual's movements and those of similar shoppers. Then we looked at how they responded to different kinds of mobile advertisements, depending on what size group they were in (alone, with one other person or with two other people).

Understanding the Impact of Social Context

Making an experiment like this both successful and robust requires several ingredients, but the overarching requirements are volume and variety of data. In other words, we will need a place where we will find thousands of people of all ages and backgrounds, who have hundreds of stores to choose from. A shopping mall is an ideal venue, so in April 2015 we partnered with the same shopping malls in Asia to collect data from several sources, including smartphone sensors we could track using Wi-Fi. As you might recall, the mall has over 300 stores spread over 1.3 million square feet. It draws an average of 100,000 visitors on an ordinary weekday and more than 200,000 visitors on weekends and holidays.

Tracking the movements of several thousand people every day gave us the volume of data to see a wide variety of shopping patterns. We received 52,500 unique user responses from 252 stores for a 21-day period. We fed the data into pattern recognition algorithms to detect—automatically and in real time—the social context of the consumers (e.g., alone, in a group of two, in a group of three, etc.) as they navigated the stores in the mall. These data helped us draw inferences about each person's level of interest in a certain store or a category of products. For example, a longer stay with slower movement with a store indicates a higher level of interest than a "hit and run" pass through a store that did not result in a purchase. Finally, when the consumer left the mall, we conducted a short follow-up survey asking additional personal information (such as whether they were a first-time visitor or not, whether they came to shop alone or with others, whether they were single or couple, whether they came to shop with family or friends, how many family members or how many friends, etc.). It was the combination of the survey data with the machine-learning-based data from Wi-Fi that enabled us to definitively pinpoint the social group dynamics for any individual user.

To decide who to target—and with what kind of coupon—we tried to find groups of trajectories which had as much as possible in common with each other, but were as different as possible from everyone else. This approach has been used in earlier experiments.[13]

We wanted to see how consumers responded to different mobile advertisements from stores they are interested in. We also wanted to study this mobile targeting impact based on their social context. We expected to see distinct differences depending on the size and nature of each group. To understand how people would respond to mobile offers based on their social context, we divided each day's participants into four groups and sent them an appropriate mobile promotion. The groups were as follows:

- "singles": consumers shopping alone
- "doubles": consumers shopping in pairs
- "triples": consumers in groups of three
- "co-located": consumers who are with other interested consumers in the same place, but not really in a social group.

To control for the potential bias introduced by the stores and products, we randomized the participation among 252 stores in the shopping mall from various categories including fashion, dining, supermarket, and so on. The consumers received the mobile coupons via text. That coupon could be based on their current location, on their trajectory, or sent randomly. For a given store, we randomized the level of price discount (e.g., 10 percent off, 20 percent off, 30 percent off, or 50 percent off). For the same level of price discount, we also randomized the coupon format (e.g., "price 50 percent off" vs. "buy one get one free")

Is it worth knowing a consumer's social context?

This chapter opened with a fundamental question for retailers, brand managers, and anyone who wants to reach a smartphone user with an advertisement: Is knowing someone's social context merely interesting, or is it very valuable?

In my own interpretation of the results you'll see in this subsection, the social context is not only valuable; it is essential. I admit I was surprised at how much of a difference social context makes in general, and also by the sharp differences across social contexts. In short, my colleagues and I expected a few differences. Instead, we saw multiple differences.

New considerations: Group composition matters

- Consumers in groups are *twice as likely* to respond to a mobile ad. Consumers respond differently when shopping alone than when shopping with in groups (irrespective of size). On average, a consumer shopping with other consumers is 1.97 times more responsive to mobile ads than a consumer shopping alone.
- "Triples" are better than "doubles." On average, a consumer in a group of three is 1.46 times more responsive to mobile ads than a consumer shopping with another consumer. The larger the group is, the better these ads perform.
- Couples may ignore you. Couples are least responsive to mobile ads on average. This finding seems to indicate that couples have some sort of attention-deficit disorder for mobile ads. Or maybe it would be considered rude to interrupt a one-on-one social interaction to check your phone.
- Offer coupons that friends can also use. A social discount coupon (such as "buy one get one free") works more effectively than an individual price

discount (such as "50 percent off") when the group contains only adults. Moreover, a social discount coupon is especially effective for groups that contain couples, but less effective for families shopping together with younger children.

- Look for wealthy "singles." High-income customers are more likely to respond to mobile ads when shopping *alone* than when shopping with others in social groups. High-income customers are sensitive to the real-time social contexts when receiving mobile ads. If your goal is to reach this audience, you have your best chance when they are shopping alone, and when you use trajectory-based ads.

Before we dig even deeper into the findings, let's see if we can reach some conclusions already about the people gathered in the food court at the beginning of the chapter. If you recall, we had three college girls sitting together at a booth, a young mother and her toddler son, a Millennial couple, and two well-dressed middle-aged men who are chatting but don't appear to be acquaintances.

It seems as if targeting the college girls with a mobile ad has the best chances of success, and targeting the very-much-in-love couple has the lowest chances. The same may apply to the other couple (Tom and Rachel) while walking the dog in Nashville, though their situation raises some questions for further research on social context: are Tom and Rachel merely a "double" when they are walking the dog, and only become a "couple" when they go out for dinner later? It would help Starbucks (their presumed destination) and Chipotle (one of many dinner possibilities for later on) to know this.

Overall what we see is that consumers are more likely to shop in groups than shop alone. But there are some subtle differences when you break the data down further. High-income consumers and older consumers are less likely to shop in larger groups (i.e., as "triples"). High-income consumers actually like to shop alone. Male consumers are more likely to shop alone and are less likely to shop with friends or children than female consumers. The implications of income levels are interesting. In fact, contrary to our general finding, the odds that high-income consumers will respond to offers *go down* as the size of their social group increases. It is a challenge to get them to bite.

So what explains this increased effectiveness of messages that factor in social dynamics? One explanation is that when trajectory-based data on users are combined with data on their real-time social dynamics, the resultant advertisement is perceived as being significantly higher in quality than a relatively simpler location-based ad and especially a random ad that lack information on users' social or group context. Advertisements based on social dynamics come across more like a personal and intimate message from a concierge than an unsolicited or irrelevant advertisement.

New considerations: Digging deeper into consumers' personalities

Next I am going to paint a futuristic picture of where brands can take this force.

When someone goes shopping with friends or family members, it unlikely that every member of that group has the same personality. Some members may be more outgoing and extroverted. Some members are more introverted. Can latent personality similarities between members of a group accentuate or attenuate the effectiveness of marketing recommendations? My work with Chris Forman and Batia Wiesenfeld had shown that the effectiveness of digital word-of-mouth recommendations depends on various characteristics of the users who create that content.

The ability of businesses to observe digital recommendations from individual users offers an attractive opportunity to learn how users' characteristics, such as personality traits, can facilitate or confine the effects of those recommendations. This is a particularly interesting question as personality characteristics have been found to affect various aspects of individual behavior such as whether people would be more likely to accept a suggested product or service,[14] preferences for music,[15] and consumers' brand preferences.[16] In the digital context, researchers[17] have found that people with specific personality characteristics, such as high openness and low neuroticism, respond more favorably to targeted advertising.

The study of personality has led to the emergence of personality psychology, which has been an identifiable discipline in social sciences for decades. A large number of researchers in this area have investigated personality constructs in an effort to uncover the underlying factors of personality, leading to taxonomies of personality traits. The most influential taxonomy of personality attributes is admittedly the "Big Five" taxonomy, and it serves as a useful integrative framework for thinking about individual differences at a fairly high level of abstraction.[18]

The five latent personality dimensions can be generally defined as follows: The first dimension is titled *agreeableness* and captures a person's tendency to be compassionate and cooperative toward others. Agreeableness is associated with altruism, cooperation, trustfulness, empathy, and compliance. The second dimension is *conscientiousness* and describes a person's tendency to act in an organized or thoughtful way. Individuals characterized by high levels of conscientiousness tend to be driven, deliberate, organized, persistent, and self-assured. The third dimension is the *extraversion* dimension and refers to a person's tendency to seek simulation in the company of others. Extraversion consists of outgoingness, sociability, assertiveness, and excitement-seeking behaviors. The fourth dimension is *emotional range*, which describes the extent to which a person's emotions are sensitive to the individual's environment. The tendency of an individual to be worried, depressed, self-conscious, and hedonistic is captured by the aforementioned dimension. Finally, the fifth dimension is that of *openness*, which refers to the extent to which a person is open to experiencing a variety of activities. Adventurousness, intellect, creativity, and liberalism define openness to experience.

While leveraging personality characteristics constitutes a promising pathway toward understanding online consumers' behaviors, there exist significant challenges that have prevented so far the exploitation of personality characteristics. Such challenges emerge from the inherent difficulty of identifying and measuring latent personality characteristics. For instance, the traditional way of measuring personality characteristic requires the completion of long questionnaires. Hence, it has been particularly burdensome, if not impossible, to obtain such information at a large scale.

Toward alleviating this problem, Panagiotis Adamopoulos, Vilma Todri, and I recently developed a method to automatically infer personality characteristics from the unstructured social media data sources such as Twitter.[19] An important aspect of our work is that our methods are simple, very generalizable across industries and can be done at a very low computational cost. All one needs to create a highly accurate personality profile of each individual user is approximately 4,000 words of social media content generated by that user![20]

This is only the beginning. I predict that similar techniques will be applied by businesses in the context of the mobile economy. As deep learning and text mining techniques get more sophisticated, such personality attributes can be identified from users' social media profiles with even more precision and ease.

In the not so distant future, I predict firms will be mining micro-facial expressions from closed circuit video cameras to infer such personality traits. By mining the same consumer's facial expressions repeatedly over time, a brand can come up with a probabilistic estimate of his or her personality, at least during that moment. This is not science fiction. Already, retailers are using software from firms like Emotient in its security cameras to gauge whether shoppers are pleased when looking at products and leaving the store.[21] Kairos is yet another firm that has produced software that analyzes facial movements such as smiles, frowns, anger, and surprise to determine feelings—translating into "a specific emotional readout" for retailers.[22] Over time, when businesses have more data on the same users from their repeat visits, they can go from mining facial expressions in order to infer real-time emotions to inferring long-term personality analyses.

What does this all imply? Well, when firms target users with mobile offers, in addition to the size and composition of the group, they can also incorporate the personality attributes of group members in their marketing campaigns to further enhance their effectiveness. Well, the next time Tom and Rachel go shopping with a friend or a relative, the marketer trying to send them a relevant offer would benefit from taking into account the three different personalities involved in the group.

Food for Thought

This is only the beginning. Stepping back from the actual details of this experiment, I believe that my colleagues and I also demonstrated a very important insight about how the proliferation of smartphone and the rise of the "tap and swipe" lifestyle will reshape business strategies at many different levels. We now have even greater confidence that these kinds of studies matter, for a number of reasons. First, they uncover differences in shopping behavior that are not immediately obvious but which have very material financial implications. Second, they show the kinds of cause-and-effect relationships, which businesses can act on. Finally, we believe we have offered first glimpse into the power of studies, which combine deep, real-time data analyses with social science. This study was the first to connect consumers' historical offline behavior (the "where we've been") with their geo-location ("where we are"), their offline social dynamics ("whom we are with") and their economic behavior and preferences. Take any one of those factors out of the decision making and the chances that a consumer will respond to a mobile advertisement drop significantly. You would be fighting a very winnable 21st-century battle with outdated, underpowered 20th-century tools and thinking.

There is little doubt that the people who are near us and around us shape our behaviors. As the race for data intensifies, organizations will be able to infer not only who we are, but also who we are with during our shopping trips. This is part of the process of firms' trying to get to know us better to create value for us. For consumers, it will pay to embrace this new phenomenon and endeavor to use it to their advantage. They can extract the value that firms will create for them based on access to their data.

Takeaways for Firms

Businesses and marketers risk hitting dead spots rather than jackpots if they design their mobile strategies without taking real-time social contexts into account. There is tremendous value in leveraging mobile devices and sensor technologies to digitize, measure, understand, and predict individuals' group dynamics in the physical environment in order to improve their digital experiences and business marketing strategies. The challenges that existed in obtaining individuals' real-time social contexts are now being diminished by the data generated from smartphones. This challenge will diminish further as consumers become more willing, even eager, to share their real-time and historical data in return for curated, customized advertisements and superior customer service. The key takeaways are the following:

• In order to tap into the full potential from the mobile economy, it is critical for businesses to take the social dynamics of their target customers in real time into account. Individual behavior varies greatly, depending on whether they are shopping alone or with someone else.
• If they are shopping with others, it matters greatly whether they are a couple, whether they are with friends or whether they are with family. If they are with friends or with family, it also matters how large the group is.

- A social discount coupon (such as "buy one get one free") works more effectively than an individual price discount (such as "50 percent off") when the group contains only adults. A social discount coupon is especially effective for groups that contain couples, but less effective for families shopping together with younger children.
- When targeting users with mobile offers, in addition to the size and demographic composition of the group, incorporating the personality attributes of group members in curating marketing campaigns can be remarkably effective.

11 Weather: Creating the Perfect Storm

In 1999 Coca-Cola conducted one of the earliest experiments with dynamic pricing for consumer products. Dynamic pricing has served as the primary system for airlines and hotels for many years. Their prices change daily, sometimes hourly, as anyone who has tried to book travel is very familiar with. We also see dynamic pricing in action when a retail store or a roadside stand tries to get rid of perishable products such as fruit and vegetables before they become unsellable.

But applying it to consumer products was unchartered territory. The soft drink giant wanted to introduce temperature-sensitive vending machines that could raise the prices for a can or bottle of soda in lockstep with rising temperatures. One industry executive criticized the initiative by saying "What's next? A machine that X-rays people's pockets to find out how much change they have and raises the price accordingly."[1] Coca-Cola eventually gave up the plan because too many consumers considered it a rip-off.[2]

But these failures didn't discourage others from experimenting with dynamic pricing, or its simpler cousin, variable pricing. In the early 2000s, the Toronto Blue Jays, a major-league baseball team, undertook a study that was unusual at the time. The team's management wanted to use data analysis to figure out why fans came to a particular game. The prevailing wisdom in baseball, and sports in general, was that only two things would guarantee that fans would flock to the stadiums in large numbers: more wins and better weather.

The Blue Jays' management thought that a more thorough analysis would yield enough insights to allow them to a better job at setting ticket prices for the next season. They would solve the puzzle that any organization faces when it sells tickets to an event: how do you increase attendance and increase revenue at the same time?

Back then a sports team, a museum, or a theater would set fixed ticket prices in advance. The more sophisticated ones would vary those advance prices on the basis of a number of factors, such as the quality of the show (or in sports, the strength of the opponent) or a fan's perceived desire to attend because of weather, day of the week, or season. The Blue Jays looked at several years' worth of data on every individual home game to try to pinpoint these attendance drivers. They were not surprised to learn that attendance was measurably higher, even adjusting for other factors, when the dome to their baseball stadium was open than when it was closed due to inclement weather.[3]

That represented the kind of "take it for granted" attitude businesses have traditionally held toward weather. You can't control it, you can't manage it, and any conclusions are rather obvious. And as Coca-Cola learned, trying to capitalize on weather variation can backfire.

Fast forward to 2016, and dynamic pricing has become much more common in the ticketing and consumer businesses. Dozens of professional sports teams and other ticket-based business now adjust their prices in real-time in response to prevailing circumstances—good or bad—including weather conditions.[4] In general, you offer fans an incentive to attend under adverse conditions: weak opponent, school night, bad weather. Conversely, you offer lower incentives or even raise prices when the weather is warm, because people are more likely to come to the ballpark and be in a better mood. One study of attendance at minor league baseball games showed that attendance fell by 136 fans for every degree below 65°F, and decreased by 26 fans for every degree of above 90°F.[5] Between 65°F and 90°F was the sweet spot for fans to feel comfortable. Such analyses offer not only guidance, but also reassurance that weather-based marketing decisions makes sense.

Conclusions such as that one, even with some precise math to back them up, still seem rather obvious, though, don't they? As the author of that study on minor league baseball admitted, you could plot the effects of weather on attendance "blindfolded without looking at any data." Who doesn't head outside and look for outdoor activities when the sun is shining and the temperatures are warm?

Then there are also some fascinating truths about how weather drives our device usage. We know that people watch less television during the summer for a number of reasons, including the fact that they are more likely to be on their feet.[6] A study based on data from American and Canadian consumers showed that tablet activity increases during the winter afternoon hours vs. summer hours. However, in the late evening and early morning hours, summer tablet activity exceeds winter activity at the same time of day.[7]

I bring up these examples to underscore the bias that people initially have when they try to factor weather into their marketing or advertising strategies. It does indeed seem at first glance that any useful insights about weather are obvious and do not warrant more detailed investigation.

In the remainder of this chapter, I will introduce a mix of academic studies and business success stories, which show that the opposite is true. The value of weather as a very influential driver of behavior is enormous and anything but obvious. In fact, weather is sometimes the most powerful of all the forces we will have looked at in part II.

Does Weather Affect Our Moods?

Weather is ubiquitous and omnipresent. Scientific literature tells us just how strongly weather can influence our behavior, our moods, and our short-term, medium-term, and long-term decision making. One study showed that when our moods change, weather can account for as much as 40 percent of that change.[8] The impact of weather on individual behavior has also been intensively analyzed in finance research. In 2003, David Hirshleifer and Tyler Shumway had demonstrated a relationship between sunshine (which they interpret as a proxy for mood) and stock returns.[9] Weather is localized and highly variable. Weather also offers that special ingredient that helps improve our understanding and find ways to make advertising a lucrative win-win for businesses and customers: lots and lots of data!

New considerations: How weather affects mobile purchases

To give you a sense of the power of weather on our behavior and our choices, I'll stick to its most basic forms: temperature, sunshine, and precipitation. A creative recent study by a team of researchers at Temple University and Fudan University[10] looked at whether there is a direct link between weather and the effectiveness of mobile advertising. The study involved sending 38 different, randomized text messages sent to more than 10 million mobile users.

The researchers isolated the effects of weather after controlling for the characteristics of the geographic locations. This clever experimental design was critical, because the effect of weather can be highly correlated with location. One would expect certain types of weather to emerge based on the combination of latitude and altitude, and therefore any researcher would have to account for that potential bias. The researchers sent text messages promoting a video-streaming service deal. Earlier researchers had asserted that people with in a bad mood are more responsive to messages with prevention framing whereas people in a good mood are less responsive to messages with the prevention framing.[11] With that in mind, one group of treated consumers got a text message that contained a prevention-focused message frame ("do not miss the opportunity to take advantage of this deal") known to elicit responses from risk-aversive individuals. The control group got text messages containing only the baseline promotion, without such framing. Weather variables recorded at the customer location at the time the mobile promotion was being sent was mapped with the above advertisement framing.

What did they find?

• Purchase likelihood increased by 31 percent in sunny weather and declined by 9 percent in wet weather, using cloudy weather as the baseline.
• Time to purchase from receipt of a mobile ad is 41.93 percent faster (49.87 percent slower) in sunny (rainy) weather.
• Compared with non-prevention ads, prevention-framed mobile ads are indeed less effective on sunny days, but more effective on rainy days. This supports the mood-risk account of the weather effects. As such, with appropriate mobile ad framing businesses can substantially benefit from weather effects.
• The results are robust to different measures for weather, and the study did account for weekday vs. weekend conditions and hazardous conditions.

Explaining this behavior

Let's think about the main result again. The study found an apparent contradiction in our shopping behavior in good weather as opposed to bad weather. One would expect that sunny warm weather would draw more people into physical retail stores, at the expense of purchase with a mobile device. The researchers found no substitution effect; in fact, purchases via mobile phones *rose* as the weather improved!

What explains this behavior?

The researchers assert that natural sunlight provides nutrients essential to short-term happiness and that, by and large, people respond positively to sunlight, thereby become more risk tolerant. An improvement in people's moods thus increases purchases on mobile devices. The fact that mobile devices are always with us makes it plausible for firms to factor in the local weather and send us offers on the go.

Another recent study by a team of researchers in Germany showed that location, time, and weather all interact to influence a potential customer's search behavior on a mobile device. Specifically, they showed that relative humidity negatively affects users' coupon choice probability and this is particularly true for the months in summer and spring. What does this imply?

- Users prefer a moderate relative humidity when they are browsing and choosing location-based coupons.[12] Going out to shop in sticky, humid weather is not desirable to most people, no matter how good the offer is.

On the basis of results such as these, it comes as no surprise that Facebook, Google, and Twitter have begun to explore how to apply weather data to their advertising strategies.

Even when we do take a closer look at the data, we still should avoid drawing broad conclusions. Weather is regional, and people react differently across the country. Hot weather in New York drives a higher click-through rate, but similarly high temperatures on the West Coast of the United States have the opposite effect on click-through rates, according to the mobile ad network AdTheorent.[13] Businesses need to take such differences into accounting when designing and serving mobile ads.

"Weather Is Our Ally": Applications across Industries

Tom Jenen, Google's former head of marketplace development and now the president of the advertising solutions provider Polar, once said: "Get to understand your customer and their behavior with weather. Weather is one of those things where you can show the usefulness of the product advertised."[14]

In the United Kingdom, the advertising team for the brewery Stella Artois refers to weather as "our ally." They began using targeted out-of-home advertisements and could trace a noticeable increase in sales for their Cidre product when the temperature was 2°C (about 4°F) above the seasonal average. They looked at 12 years' official weather data and saw that the immediate prevailing conditions, not seasons or seasonal averages, had a significant impact on their sales.[15] The ads offered refreshment when the need for refreshment had increased.

A major chain of quick-service restaurants used weather in a very creative way to reach mobile audiences near its physical stores. It wanted to promote its selection of hot soups, sandwiches, and salads during lunch time. Working with its agency partners, it created mobile ads with meal options based on local weather conditions. When temperatures dropped below 60°F, the ad creative promoted the restaurant's hot selection of fresh soups. If the temperatures were above 60°F, the ad creative would focus around the brand's salads and sandwiches. Weather events such as snow or rain also triggered the soups-focused ad creative. The campaign exceeded the client's benchmarks by 124 percent and generated an above average click through-rate (CTR). Perhaps most importantly, it also caused a measurable sales lift in the targeted markets.[16]

Procter & Gamble (P&G) is another company that has taken the link between advertising and weather to a very personal level in order to demonstrate the usefulness of its products. The shampoo category is very crowded, and Procter & Gamble wanted to help boost sales at Walgreen's, one of its key retail partners, who was struggling to improve its performance. P&G's Pantene line of hair care products was also losing market share overall. How could the manufacturer and the retail chain reach consumers in a targeted way that also showcased the product? They linked Pantene to weather. The idea was that one decision women make every day is how to do their hair, and the weather forecast—humidity, wind, temperature—plays a role in that decision. The critical step P&G took was to create the "Haircast," a weather forecast that the target audience (women) could call up anytime on any given day. The video would describe the local weather in terms that can affect hair styling, and then would recommend the appropriate Pantene product for that situation.[17] Without introducing any new products, P&G and Walgreen's cooperated to boost sales by 10 percent over a two-month period, and increased Walgreen's overall sales of hair care products by 4 percent. Understanding how the consumer used the product based on weather conditions made the difference.

Costa, a chain of coffee shops in the UK, partnered with The Weather Channel to promote its summer drinks, with targeted advertising activity running online when the temperature hit 22°C (about 72°F). This resulted in a click-through rate of 0.15 percent on a homepage takeover, 0.52 percent on an iPad, 2.72 percent on a mobile banner and 0.47 percent for a display ad.[18]

Using data made available by Weather Co., the arts and crafts retailer Michaels likewise uncovered a data-driven insight that would allow it to design and customize very specific localized promotions. It learned that purchases of arts and crafts supplies tend to peak three days ahead of a forecasted rainy day.[19]

Even Coca-Cola has caught up with the times. A little more than a decade after its first foray into dynamic pricing, Coca-Cola launched temperature-sensitive vending machines again, this time in a Spain. The difference is that these new machines offered a *discount* as the temperature increased. A can would cost 2 euros up to 25°C, but would cost only 1 euro if the temperature rose above 30°C (86°F).[20]

Forecasting sales of consumer products and services with some precision has clear advantages. But these are relatively small outlays. The price of an alcoholic beverage probably ranges between $5 and $10. Even in more extravagant cases—a box seat at a ball game, a restaurant bill for you and a few guests—the expenditure is probably in the range of $100. When inclement weather compels Uber to introduce surge pricing, its version of dynamic pricing, the added cost still won't bankrupt anyone.

It would be interesting to see whether the weather has similar effects on our behavior when we want to buy a durable good, such as a car or a house. Besides a college education (at least in the United States), buying a house or a car is among the largest purchases someone will ever make in his or her lifetime. One would expect a considerable amount of planning and forethought behind such purchases, which would offset or negate the effects of the weather on any given day. No one is going to make such an expensive decision with far-reaching consequences on the basis of the weather that day, right?

Wrong.

New consideration: Weather drives sales of big-ticket items

One study from the National Bureau of Economic Research found that sales of black vehicles are sensitive to both temperature and sunshine. If the temperature rises by 20°F, sales of black vehicles fall by 2.1 percent. If the weather is completely clear, sales of black vehicles fall by 5.6 percent compared with sales on cloudy days.[21] The researchers also found that warmer days increase sales of convertibles, up until the warmest days of the year. At that point, an increase in temperature probably makes driving a convertible seem less appealing, whereas unseasonably warm days during cooler months make buying a convertible seem like a great idea. Another interesting finding was that clear skies also increased convertible sales by the same amount as an increase in temperature of 16°F. Temperature also influences the price someone would pay for a house with a swimming pool. A house sold in a month with an average temperature of 80°F would sell for 0.65 percentage points more than the same house sold in a month with an average temperature of 30°F.

A New Consideration: Micro-Climate as a Treasure Trove

Think back to Coca-Cola. The source of the data for their price changes is the vending machine itself. It is hard to get more local than that. But if an advertiser doesn't have a weather beacon on every corner, how can it understand the immediate conditions as consumers' faces, and design a targeted ad on that basis?

When we check the weather on our smartphones, we tend to type in a city, but usually receive maps that show a region, often spanning multiple states. You may see a 40 percent chance of rain for New York City on average, but a 90 percent chance in the morning. You will also get a temperature forecast of, say, 75°F that day. At the same time, anyone who lives in a climate with distinct seasons and active weather patterns knows that a few miles can make a huge difference. The temperature can vary noticeably between Newark Airport (in New Jersey, about 15 miles southwest of Midtown Manhattan, and John F. Kennedy Airport (in the borough of Queens). The variation can be even more extreme. The lake-effect snowstorm that hit Buffalo, New York in November 2014 dumped more than 5 feet of snow on the city itself, but only "mere inches" a few miles to the north.[22]

How can a business gather information about such discrepancies and take advantage of them in real time? One increasingly common way is to mine data from what Weather Co. (formerly The Weather Channel) refers to as microclimates. Imagine the United States divided up into a grid with each square roughly 1.2 miles by 1.2 miles. That is exactly what scientists at the Oregon State University did over a decade ago, when they took on Weather Co. as a partner. They created a map of the United States with around 3 million separate "micro-climates," each with the potential to have its own unique real-time weather profile.[23] For example, the city of Chicago covers roughly 230 square miles. This grid would divide the city up into about 160 separate weather zones. Anyone who has traversed the city, or looked out the window from a rooftop or a high-rise window, knows that even within Chicago, everything from sunshine to snowfall can show a wide variance depending on where you are.

If a business wants to target someone at a specific location in Chicago at a certain time of day, it makes no sense to overlay a generic forecast for Chicago (75°F with 40 percent chance of rain) on that model. The weather at a given location could be 65°F and cloudy with no precipitation all day, despite what the "Chicago" forecast said. Another location on that same day could be 81°F with the nuisance of light showers almost all day long. Knowing that information would help sharpen decisions about whom to target, with what product and with what incentive.

"Depending on the city you are in or the microclimate you live in," Vikram Somaya of Weather Co.'s WeatherFX Division told the *Wall Street Journal*, "your relationship with products is different." In that same article, Paul Walsh, Weather Co.'s vice president for weather analytics, said: "The old paradigm of business and weather was cope and avoid. With technology, the paradigm is now anticipate and exploit."[24] Weather Co. has more than 75 years' worth of information on temperatures, dew points, cloud-cover percentages, and many other phenomena for much of North America and also for other territories.

Micro-climates should allow for micro-targeting of consumers, regardless of whether the products have a direct link to weather conditions. The marketing solutions provider Weather Unlocked offers some guidance on what works and what doesn't in terms of advertisements.[25] The first one sounds very intuitive: match the product closely to the conditions, i.e., sunscreen on hot, sunny days and warm coats on snowy days. Speaking of cold, some of their findings illustrate some of the stronger contextual power of weather. Being cold can increase our desire to seek warmth and comfort, which can be psychological and not only physical. One study showed that ticket sales for romantic comedies increase during cold or wet weather. This may be due in part to the content of the films and also because events such as films, concerts, or restaurant visits are group activities, which can fulfill that need for warm and comfort beyond that ad for that warm winter coat. The same study also showed that when people are exposed to the cold, they seek "social" activities and buy products that engender a feeling of psychological warmth. Examples include events that have an element of group participation, such as tickets for films, gigs, theater or restaurant bookings, and so on.

As I have done throughout the book, I will again warn against the temptation to draw blanket conclusions about any cause-and-effect relationship involving mobile advertising. Weather is a special case, because we have taken a lot of "macro" effects for granted, so much so, that many business have been reluctant to get more precise data on how weather drives our personal decision making. Part of this bias may come from the fact that none of us can escape weather, and that we extrapolate the logic behind our own experiences to universal truths. Warm weather improves our mood, warm weather makes us buy certain products, and cold or rainy weather does the opposite.

Although the effect of micro-climate-based targeting has not been academically tested yet, all the indicators mentioned above suggest it can be a powerful instrument for micro-targeting.

Food for Thought

In a report on the effects of weather on mobile advertising, a team of researchers concluded that "weather is a much under-utilized environmental context variable. Sometimes the critical thing is not the cleverest ad creativity, but rather environmental forces occurring naturally, right under our noses."[26]

Put another way: There is now ample evidence that our immediate environment plays a crucial role in influencing our mindset. Weather is one of the biggest components of our immediate environment. Because of its influence on consumers' behavior, weather is one of the most accurate predictors of purchasing patterns next to macroeconomic variables.

There is an implied lesson about geo-fencing in the fact that Weather Co. breaks down the United States into 3 million micro-climates, each with its own characteristics. It is not hard to imagine that each of these micro-climates has its own unique set of economic and demographic measures as well, from traffic flows to concentration of businesses to income levels and education levels to preferences and tastes.

The available data and analytical power in 2017 and beyond far exceeds what the Blue Jays could have had at their disposal in 2003, or Coca-Cola back in 1999. We can't change weather and we may still get frustrated when predictions aren't always reliable. But increasingly we can predict its effects on how consumers behave and how responsive we will be to mobile advertisements.

To summarize, the weather affects what consumers buy, where they go, what they wear, what they eat, and how they feel. In the next few years, consumers will see more and more firms across a wide spectrum of industries making clever use of variations in the patterns of micro-climates to customize offers for us on the go. I would encourage consumers to expect this paradigm shift and embrace it.

Takeaways for Firms

Weather can strongly influence consumers' behavior, moods, and decision making. Weather certainly influences their use of mobile devices. It also affects when and to what extent consumers respond to advertising. Weather has been shown to affect purchases in food and drinks, clothing and fashion, travel and hospitality, leisure and entertainment, health and beauty, home and garden, energy, insurance, and many other industries.[27]

- Historically, weather and television viewership (and therefore impact of advertising) has been negatively correlated. But weather and the use of mobile devices are positively correlated. Purchases via smartphones increase in good (sunny and less humid) weather and decrease in bad (cloudy and more humid) weather. Also, in good (bad) weather, people take less (more) time to respond to mobile ads.

- Avoid sending negatively framed copy (e.g., "Don't miss out") on sunny days, and use such ads instead on rainy, snowy, foggy, or stormy days to boost the purchase of goods via smartphones.[28]
- Micro-targeting of consumers based on micro-climates can be a powerful instrument.
- The lift from including weather in your omni-channel marketing strategy—in conjunction with other channels such as television and the Internet and in conjunction with other forces such as location, context and time—is going to be meaningful. This is true not only for frequently bought, inexpensive products but also for big-ticket items.

12 Tech Mix: Solving Wanamaker's Riddle

Imagine the 19th-century merchant John Wanamaker standing in a long line at the airport security control on a summer weekday. He is the one who famously posed a marketing riddle: "Half the money I spend on advertising is wasted; the trouble is, I don't know which half."

Wanamaker died in 1922, when the original electronic medium (radio) was in its commercial infancy and television did not exist. Almost a century later he would see legions of people, from the very young to the very old, handling all kinds of gadgets as they move through the security lines. They juggle smartphones, tablets, e-readers, and perhaps a hand-held game device or a portable DVD player, and they pull personal computers of various shapes and sizes from their carry-on bags and lay them on the belts for scanning.

The screens of all these devices are ways to reach customers with advertisements, and as I have pointed out, investments in this market, are growing exponentially. But Wanamaker's riddle is still relevant. When advertisers spend tens of billions of dollars in aggregate to send their messages to prospective customers, how effective are their efforts? A dense fog has persisted for more than 100 years between the presumed delivery of an advertising message and the influence it has had on consumers.

Hence, the last (but certainly not the least important) force I want to discuss is the concoction of devices and ad formats. I call this concoction the "tech mix." Think of this as a force that enables businesses to pursue omni-channel marketing. It has two dimensions. First, consumers today spend time on multiple devices (or multiple screens) and this sort of multihoming creates an inter-dependency between devices that firms can tap into. Second, consumers are exposed to multiple ad messages in different ad formats for the same brand across different channels at different points in time in their path to purchase. This creates potential omni-channel synergies. These two dimensions of tech mix influence consumers' behavior in non-trivial ways and have made digital attribution the holy grail of advertising.

I will begin by talking about multi-screen behavior.

Interdependencies between Devices in the Mobile Ecosystem

With the proliferation of devices and screens per consumer, companies are wrestling with the question of how this multi-device world is shaping user behavior. The introduction of tablets in online retailing has created an additional touch point through which e-commerce companies can interact with consumers. In this vein, "mobile" does not necessarily mean "smartphone." It could also mean "tablet." In the path to purchase, some of us start our Internet search on a desktop, then move onto searching on a tablet and then eventually we may buy the product on our smartphone. Some of us do the opposite. We may start off on a smartphone before ending up buying via a desktop. See figure 12.1 for an example.

The number of tablets in use worldwide is expected to reach 1.43 billion by 2018.[1] One question that many managers have been curious about is

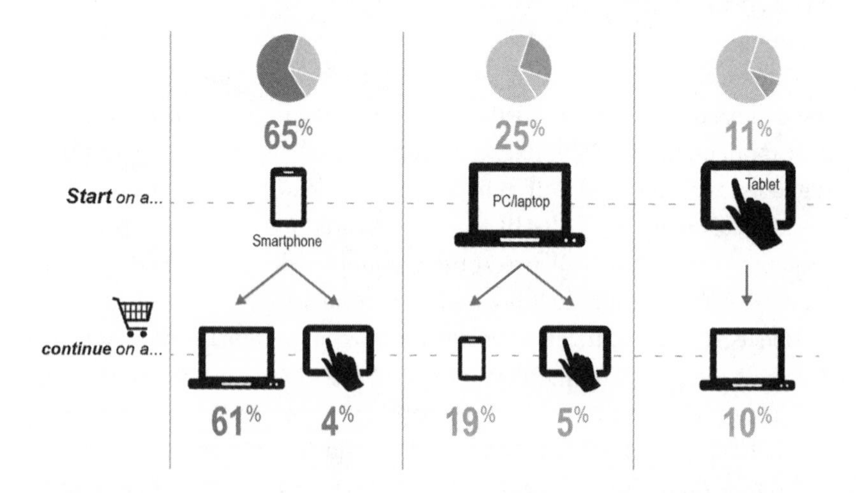

Figure 12.1
We use multiple channels when we buy goods and services online. (adapted from a figure produced by Google)

whether these multiple screens compete for attention or do they complement each other? In a recent study based on data from the Chinese e-commerce firm Alibaba, Jason Chan, Sangpil Han, Kaiquan Xu, and I examined whether the tablet acts as a complement to or a substitute for the smartphone and personal computer channels.

New considerations: Substitution and complementarities between devices

Our results demonstrate that the tablet channel acts as a substitute for the personal computer channel, but it acts as a complement for the smartphone channel. Sales through the smartphone channel increased by 55.6 percent, while the sales through the personal computer channel decreased by 14.5 percent after users began using iPads to access Alibaba. The use of tablets spurs casual browsing that leads to the purchase of more impulse products as well as a wider diversity of products, via the smartphone channel. Sales occurring through the desktop channel do go down a bit. But overall there is a growth in the revenue pie; that is, net sales increase for the ecommerce retailer.

Our findings are consistent with those of another team that analyzed data from eBay in the United States.[2] They found that the adoption of the mobile shopping application (used on smartphones and tablets) is associated with both an immediate and sustained increase in total purchasing from eBay. The data also show that mobile application purchases are incremental purchases; that is, these are simply purchases that in the absence of the mobile platform would have not happened otherwise on the regular Internet platform. Six months after the adoption of mobile channel on eBay, the average number of non-mobile purchases remained virtually the same to its pre-adoption rate, but overall purchasing doubled (on average) as a result of incremental purchasing activity via the mobile app.

In our study on Alibaba, we had also examined the temporal aspects of user browsing behaviors to understand how the introduction of the tablet channel to the existing mix of channels can affect browsing behavior in more detail.

Our analyses showed that the degree of inter-relationship between devices varies across the course of the day. The introduction of the tablet results in a significant gain in the proportion of browsing frequency share for the mobile channel during the commuting hours from 7 to 9 a.m. and from 5 to 9 p.m. The gain in browsing share during commuting hours indicates that the complementary impact of tablets works by spurring users to increase their browsing activities on their smartphones. Past the evening commute hours, the share of smartphone's browsing frequency decreases till 11 p.m., while the browsing share of tablet increases steadily, suggesting that users tend to switch to tablets for browsing when they are at home. The simultaneous decline in mobile browsing and gain in tablet browsing within a close time frame is also suggestive of the likelihood of tablet users utilizing micro-moments from earlier mobile browsing sessions to continue their shopping. Interestingly, we also saw an increase in the share of mobile browsing frequency from midnight to 3 a.m., which is aligned with the general observation of increased use of mobile devices before sleep time.

The way consumers use smartphones and tablets also varies by day of the week as well as by season.[3] Use of such devices drops off for smartphones and tablets on weekends, with a stronger effect in the summer than in the winter. The drop-off is also more pronounced for smartphones than for tablets, with summer smartphone use falling by as much as 15 percent on weekends compared with peaks on weekdays. But the general pattern on any given day for either device holds true even as overall use declines. The maximum share of use on either device in any season occurs in the early evening hours, gradually declining to reach a low in the early morning, and then rebounding in the late morning and in the afternoon.

Some consumers use the mobile device as a second screen, alongside a first screen such as a personal computer or a television. When they do that, the content on the first screen can be a contextual factor to consider in determining tactics for mobile advertising. For example, advertising on the mobile device (i.e., second screen) could target the first screen's content (e.g., a television show) or be synchronized with advertising on the first screen.[4] This can be seen as a consequence of the phenomenon of connected viewing where consumers watch television shows while simultaneously engaging with the same content on a smartphone or a tablet. Some TV networks have been aggressive about driving viewers to a second screen experience (for example, through an app such as Syfy Sync).

Eywa Media Innovations, a digital signal processing technology firm based in Singapore, uses deep learning and voice recognition technology to send audiences real-time mobile offers based on the TV shows they are watching at any moment. Imagine a person is watching his or her favorite TV show. During the show or break, when an ad comes up, the

EYWAMEDIA app on the persons phone is able to recognize the content of that TV ad or show and, in real time, send that person a targeted ad or an offer on their mobile phone for the very same brand that was in the TV advertisement; this is purely based on individual viewing patterns and preferences.[5]

As these kinds of cross-screen targeting continues to increase in the coming years, these cross-screen synergies will be a major factor driving mobile conversion rates.

Digital Attribution

The data and insights I have presented so far can help marketers move closer to solving Wanamaker's riddle. They can measure advertising effectiveness and return on investment more precisely because the huge amount of real-time data from consumers has begun to lift that dense fog. In many cases in the modern mobile ecosystem, we know immediately who interacts with an advertisement, when and where they do, and whether and how they ultimately buy a product or a service. We are closer than ever, but there is still work to do in the measurement of advertising effectiveness.

Why is it important for firms to measure audience behavior and ad effectiveness? Unless firms quantify these metrics, they will do a poor job in allocating their advertising budgets. Edward Deming's famous statement that "you can't improve something you cannot measure" remains true.

Using existing metrics does not address how much each advertising touchpoint contributes to making a sale. "First click," "Last click," and "Equal Weight" models fail to enable a full understanding of multi-channel influences on conversions and will generate misleading insights. A "first click" model assumes that all the credit for the purchase goes to the first ad that the consumer saw in his path to purchase, a "last click" model assumes that all the credit goes to the last ad in his path to purchase and an "equal weight" model assumes that all the ads the consumer saw get equal credit. Clearly they all are fundamentally flawed. We need to develop new models of attribution to facilitate the allocation of advertising dollars on the basis of the influence of specific touchpoints in the path to purchase journey on purchase behavior.

A number of other scientific studies have also tackled the digital attribution problem head on and generated fascinating insights.[6] From a big picture perspective, an appropriate digital attribution framework is one that is generalizable across multiple industries and products. It facilitates performance-based advertising by allowing advertisers to reward their digital display networks partners on the basis of their true effectiveness in driving key marketing objectives. It allows advertisers to perform real-time analysis of their advertising campaigns, which is critical as the world moves increasingly toward programmatic marketing. Finally, it also enables better optimization of granular advertising choices for advertisers, such as the effectiveness of display creative strategies or the identification of the optimal time intervals between advertising exposures.

The good news for firms is that academics are fully cognizant of the importance of this problem. The framework we have built in our paper can be effectively used by brands to identify subgroups of individuals that have been exposed to different tactical advertising strategies in order to identify their true causal effect.

New considerations: Viewability of ads

In a paper published in December 2016, Vilma Todri and I tackled several aspects of this notoriously difficult attribution problem in the context of digital advertising. We provided a set of new statistical models that facilitate causal inferences from observational data.[7] Our framework is part of a pioneering set of research focused on digital attribution, or determining which advertising touchpoint deserves the most credit for a consumer's purchase—a key question among advertising executives who manage campaigns for brands. We used an individual-level data set consisting of several million data points that had information on not only which user viewed an ad but also the duration of the ad exposure. These novel data helped us capture the effectiveness of digital advertising across a wide range of consumers' behaviors. On average, 55 percent of display ads are not viewable—meaning they are not visible on a consumer's screen area for more than one second. This makes one think about how much money is being wasted.

This non-viewability of an ad can happen due to many reasons. Several contextual factors, such as the browser window size, the screen resolution, and the screen orientation of the user might determine whether an advertising impression will be rendered viewable. Other possible scenarios include circumstances under which the viewer does not have the appropriate plug-ins for an interactive ad to be displayed, or when the viewer utilizes some type of ad blocker software—so that even if the browser loads the impression, the user never views the ad. An impression is also rendered non-viewable when the publisher places an image or another layer overlapping the ad. The *New York Times* displays such a layer when a non-subscriber attempts to view more articles than the *Times'* paywall limit allows.

We also found that consumers can be up to 28 percent more likely to engage with a display advertisement when they are targeted early in their funnel paths. By studying the dynamics of these effects, we showed that advertising effects are amplified up to four times when consumers are targeted earlier in their path to purchase.

Interdependencies in the Digital Ecosystem

Let's first talk about consumers' cross channel use behavior. It is tempting to lump Web-based advertising (personal computer) together with tablets, and smartphones under headings such as "digital" or "online," but as we will see in this chapter, each is its own world as far as engagement, conversion, and use patterns go. Marketers also need to be careful when they implement mobile advertising strategies because of the complex cross-effects with other forms of advertising. Can the effectiveness of mobile advertising be improved by combining it with Internet advertising?

To investigate this, Sung-Hyuk Park, Sangpil Han, and I undertook a large-scale, real-life experiment with a digital goods retailer who relied on both W-based and mobile advertisements. This retailer was the largest of its kind in South Korea and one of the top three retailers in the Asia Pacific region. We randomly selected the digital products and observed the click-through rates and conversion rates on a personal computer and a mobile device over a six-week period for the ads for those products. We were able to manipulate the availability of the digital goods, which were advertised and sold through both channels (Web and mobile). We had around 26 million advertising transaction records available for analysis.

Our goal was to understand the stand-alone and synergy effects across the two platforms. The various forms of advertising in the randomized field experiment were as follows:

- scenario 1: ads on Web channel only
- scenario 2: ads on both Web and mobile channels

- scenario 3: ads on mobile channel only
- scenario 4: no advertising at all.

In subsequent periods we would reverse the order in which we used the channels.

New considerations: Synergies between Web advertising and mobile advertising

The results were compelling, for click-through rates as well as for conversion. The Web click-through rate rose by 34 percent when we added mobile advertising to the mix, relative to when the advertisements appeared only on the Web. In contrast, the mobile CTR improved by 23 percent when Web advertising was added to the mix, relative to when the ads only appeared on the mobile channel. On its own, Web advertising had a higher CTR, but the important insight here for businesses allocating their advertising dollars is that an investment in a supplementary channel improved click-through rates significantly, regardless of whether we added mobile to Web or Web to mobile.

The conversion rates also showed a different result that may seem surprising at first glance. The Web conversion rate increased by 36 percent when we added mobile advertising to the mix. But the mobile conversion rate fell by 16 percent when we added Web advertising to the mix. This may indicate at first glance that the mix of mobile and Web advertising is detrimental to conversion relative to mobile-only advertising. However, this is true only when we look at the channels in isolation. Overall conversion was robust and actually reveals a more powerful finding. We found that in the latter case mobile conversion declined because the advertising combination drove higher conversion via the Web. Customers saw the advertisement on both channels, but preferred to do the actual transaction on a personal computer rather than on a smartphone. Even with the rise of mobile devices, there are still many transactions that consumers prefer to do on a computer!

These numbers provide advertisers a starting point for finding the optimal allocation of budget across the Web and mobile channels, keeping in mind that the combination of the two channels trumps the use of either in isolation. Marketers should move away from creating silos by channel. Web and mobile advertising act synergistically and affect each other in positive ways. Allocating budget for one must consider the synergy effects on the other. The mix of ads acts as a better memory cue to recall and engage. Said simply, and somewhat counter-intuitively, firms should not forget about the Web while they are wowed by mobile channels.

New considerations: What drives mobile conversion rates?

When people claim that mobile is the channel with the lowest conversion rates, they are overlooking the role mobile plays within the consumer's purchase decision process. A team of researchers led by P. K. Kannan used European retail data to show that switching from smartphone to personal computer almost doubles the conversion probability on average, compared with when a consumer continues shopping with the same device.[8] The mobile channel may have a low immediate conversion rate, but it makes a vital contribution as it aids conversion in other channels.

The study offered two other insights that are exciting not merely for their content but because they show yet again the kinds of relevant questions researchers can answer once they have access to the right kind of consumer data.

- Does a consumer's familiarity with a retailer affect mobile conversion rates? Yes, it does. The more experienced a customer is with an online retailer, the less impact device switching has.
- Does repeated viewing of the same product affect mobile conversion rates? Yes, it does. In this case, device switching leads to greater conversion, especially when the products have relatively high prices and the time between viewing sessions was short.

Differences between the Mobile and PC Channels

Regardless of the channel (mobile or personal computer), a few insights seem to be clear after two decades of commercial activity online. The Internet helps people overcome geographic isolation by offsetting or eliminating the effects of distance and reduces search costs.[9]

If you need a visual refresher of how much better we have it nowadays with the Internet, you could watch some old movies. Witness what Bob Woodward (as portrayed by Robert Redford) endures in the 1976 movie *All the President's Men*. When he wants to know whether someone at the White

House checked out a particular book from a library, a library clerk hands fills up a table of paper records bound with rubber bands. There are probably several thousand individual records on the table, and Woodward and Bernstein have to sort through the records by hand. Later in the movie, Woodward wants to track down where someone lives and how to call them back. His method is the best one available at the time: go to his company's library and review the telephone books from all area codes throughout the United States—by hand—until he finds the name. We shouldn't take for granted how far we have come from what counted as state-of-the-art research 40 years ago.

The question, then, is whether searching on a mobile device takes us a further step forward in efficiency, or whether using a smartphone is in reality less "Internet-like" than searching with a personal computer. We expected two results when my colleagues and I conducted a study to test this. First, we thought that search costs would be higher on the smartphone than on the personal computer. These are not out-of-pocket expenses, but rather the mental effort required to evaluate search results. We also anticipated that the distance from a particular store would be more important for someone browsing on a mobile device than on the personal computer.

The implications here go far beyond the technological ones, i.e., that the mobile channel is indeed less "Internet-like" than the personal computer. I truly believe (and have seen this in various data sets I have worked with) that consumers use mobile devices to shop for different kinds of products than those on the regular Internet. One such difference is whether a product is a commodity (or more commonly bought) product as opposed to a niche (less commonly bought) product. In fact, the team that worked with eBay data asserted that "mobile devices, with small screens, may be easier to use to shop for commodity items, as opposed to idiosyncratic items that require more careful inspection." Their data indeed showed this to be true.

Solving Wanamaker's Riddle: New Industry Tools

These insights really do bring us all closer to solving Wanamaker's riddle, because we can understand better than ever before what types of advertising work, monitor engagement and conversion, make specific corrections when one approach underperforms, and aim for a real optimum rather than a theoretical or assumption-driven approach.

Let's get specific. If a retailer is planning to launch a mobile ad campaign based on individual geo-targeting, the results of all the experiments in this book help determine where to erect that fence for targeting users. In chapter 5 I introduced Linda, who was standing in Manhattan and needed to buy an outfit ahead of starting her new job. It is possible for any retailer within a given radius to model the chances that Linda will click on a mobile promotion, and whether she will follow that offer with action.

But there are more ways to reach Linda than pushing an offer as a mobile promotion. What if Linda, while standing on that Manhattan street corner, decides to search on Google for all stores near her current location that stock the outfit she has in mind. Businesses can also make themselves more discoverable by investing in mobile search engine marketing, an approach that has become more critical as search activity shifts from personal computers to mobile devices.

Advertisers have often complained that they need to solve the online-to-offline attribution problem. Digital marketing does not enable them to track incremental foot traffic in their physical stores and offline conversions. For example, if a user clicks on a search engine ad and then at some point decides to walk into the store, the store would have no way of knowing (short of a survey) that the user had clicked on its search engine ad first before coming to the store premises. In response to that, Google introduced yet another magic wand. The AdWords Store Visits tool in 2014 lets advertisers track the incremental foot traffic they get from search engine marketing. According to Google's estimates, it has measured more than a billion store visits globally in the 2 years since its introduction. There are lots of high-profile success stories with AdWords Store visits. Using the AdWords Store visits tool, the auto manufacturer Nissan found that 6 percent of mobile ad clicks result in a trip to a dealership, delivering an estimated 25× return on investment.[10]

Attribution Powered by Foursquare, launched in February 2016, is another new tool that helps address the online to offline attribution problem. It draws data from a panel of 1.3 million Foursquare users who have agreed to leave their location-sharing feature on at all times, so that Foursquare knows every store they visit—even if they don't open the app.[11] It can tap into 65 million venues in 100 countries listed on Foursquare and has been tested by the restaurant chain TGI Fridays, the spirits giant Brown-Forman, the publishing app Flipboard, and many other firms.

Foursquare wanted to explore how it could eventually be tied back to TV advertising and attribution. It analyzed foot traffic to the stores of Super Bowl 50 advertisers in the week after the game. The assumption was that consumers involved in their study watched the Big Game commercials and compared their visits in the week after the game with those in the preceding weeks. Red Lobster saw a 12 percent increase and Taco Bell a 10 percent increase. In the automotive sector, Hyundai got a 5 percent increase in foot traffic.[12] According to Foursquare's president, Steven Rosenblatt, Foursquare can provide these data to advertisers daily rather than making them wait for weeks, and it has years of location data that can be used to ascertain whether or not a given consumer really is in a certain location.[13]

In April 2016, YouTube partnered with Oracle to track in-store sales of consumer packaged goods on the basis of views of YouTube's TrueView video ads that play before a viewer's selected video. In June 2016, Facebook announced its Offline Conversion API tool that lets retailers partner with a number of in-store systems companies including Square, IBM, Marketo, and Lightspeed to track in-store traffic and measure if users bought something that was recently advertised in their area in the digital domain.[14] By virtue of including a native store locator for mobile ads, Facebook is also enabling advertisers to more accurately measure their campaigns' effect on in-store sales with new features and metrics.[15]

Facebook now provides free app analytics to advertisers.[16] It makes app analytics and cross-channel attribution very accessible. The premise is that people use Facebook on many devices, so if Web and app developers use this platform then they can track people all through the funnel across multiple devices. Suppose a business has a cross-channel presence and it wants to retarget people who abandoned its online shopping carts by serving them a relevant ad on Facebook on their Android phone while they are walking past a competitor's store. Or say a music app wants to retarget people on the basis of their in-app behavior. This new toolkit will go a long way in enabling these kinds of analytics. We can think of this new tool as a mobile version of Google analytics as it links broad app behavior to Facebook-based advertising. It seems like a natural outcome from the use of Atlas, the ad server that leverages Facebook ID for cross-channel attribution and allows advertisers to use Facebook's data about its users to target on non-Facebook apps and websites.

The Significance of the Mobile Ad Format

Tech mix or omni-channel doesn't only mean hardware. How we present a digital advertisement also determines its effectiveness. Advertisements still present a challenge on the creative side. The content of the ads themselves requires careful optimization. Mobile advertisements are delivered through a channel filled with a lot of other content, and people are often inclined to pay minimal attention. For an advertiser who wants to design a campaign for a specific goal and target audience, at the time of writing this book, there were several choices: banner ads, in which the ad appears alongside the content of the site; native ads, in which the ad appears within the content; video ads, in which the ad appears within (mid-roll) or before the content (pre-roll) in a video; and interstitial ads, in which the ads fill the whole screen. I fully expect new ad formats to evolve as technology advances. As advertisers take them on, there will be more testing that will be needed.

Which form works best? Which kinds of ads should be within apps and which ones should be on mobile optimized websites? There are also potential negative synergies due to redundancy of information and inefficiency if the messages aren't specific to the channel. The user interface also matters a lot in mobile. According to a study conducted by Google, advertisers need to provide consumers with what they are looking for right away in mobile: faster loading, large buttons, easy searches, and limited scrolling.

The format of the ad and the richness of the content embedded in it also plays a big role in influencing customer engagement and impact in the path to purchase.[17] The differences are significant enough to compel marketers to pay more attention. Success of ad effectiveness will depend on format and operating system. On Android, banner ads delivered the best conversions for lifestyle content (2.02 percent), according to InMobi. On iOS, banner ads worked best for classified content (2.7 percent). If you want to use video on mobile, you get higher play rates in-app (14 percent) than on mobile Web (8.3 percent). The same pattern holds true for expandable ads. Interstitial ads work best with games. Amobee is one of several firms that are helping advertisers navigate a cluttered mobile environment. Its "Out of the Box" ad format offers a customized, animated introduction that expands beyond the standard banner size to capture viewers' attention and encourage engagement. Clients such as Lexus, PayPal, and Red Bull have successfully utilized these kinds of tools for their mobile campaigns.[18]

The Big Potential of Native Advertising

Despite these initiatives, there is no question that many consumers are skipping ads or even completely blocking them. Pop-up and banner ads that interrupt our mobile Web experience can be frustrating. Apple's embracement of ad blockers in iOS 9 is a consequence of this kind of frustration among consumers. The adoption of ad blockers and the backlash over excessive advertising by consumers is making firms think about creative ways to circumvent the problem.

In this regard, native ads that are blurring the lines between advertisement and content are proving to be remarkably useful.[19] An advertorial is the simplest version of a native ad. Other simple examples include Twitter's promoted Tweets, Facebook's promoted stories, and Tumblr's promoted posts. We will increasingly see publishers and advertisers working together to produce native ads. Native ads have an inherent advantage because they leverage the context of the content where they appear. The data support the idea that this would increase user engagement, in terms of both views and shares. Native ads are viewed 53 percent more than banner ads and in fact, 32 percent of consumers said they would share a native ad with friends and family, compared with 19 percent for banner ads. The reach can be staggering. Using video ads on Facebook, Heineken Light reached 54 percent of its audience—35 million people—in just three days.[20] Coldwell Banker Real Estate LLC and other advertisers that are moving money out of display ad and embracing native ads are experiencing a 50 percent higher click-through rates as these native ads blend seamlessly into the user's content feed on many websites.[21]

Regulators have often expressed concerns that, because native advertising mimics non-sponsored content on the same medium, it can be confusing to many consumers—or even deceptive when the advertising disclosure is incomplete or inappropriate.

A new research study now dispels this misconception. In a large-scale field experiment involving mobile search for more than 600 restaurants on a sample of 200,000 consumers, two researchers, Navdeep Sahni and Harikesh Nair have demonstrated that transparency of ads in the context of mobile search can greatly enhance consumers' evaluations of advertised goods. They examined this in the context of Zomato, an India-based mobile search platform that provides listings and reviews for restaurants for consumers to search and browse. The experiment randomizes users into a control condition in which no ads are shown and two treatment conditions in which users see ads for local restaurants. Both treatment conditions show the same ads, but differ in the manner in which advertising is disclosed. Consumers in the first group are shown the advertiser's listing without any disclosure, whereas consumers in the second group are shown the listing with the disclosure that the listing is an ad.

These researchers found that disclosure of a listing as an advertisement alone increases calls to the restaurant by 77 percent. This effect is higher when the consumer uses the platform away from his typical city of search. In such cities, because of lack of familiarity, they are less certain about the quality of restaurants. This effect is also higher for restaurants for which consumers *a priori* have greater uncertainty such as those that have received fewer ratings in the past. The important takeaway is that both consumers and firms benefit from such native ad disclosures. When the listings are disclosed as ads, consumers choose restaurants that are better rated and in the same vein, advertisers gain from the improved conversion induced by disclosure. Overall, this study found no support that native advertising tricks users and drives them to advertisers.

Most native ads today leave a lot to be desired. But as one observer said, "when it is done with as much flair, relevance and journalistic integrity the WSJ/Netflix effort, it is a beautiful thing to see, which readers want to read and which ad blockers have no interest in blocking."[22]

Choosing between an App and a Mobile Website

Should businesses invest in a mobile app or in a mobile optimized website or both? This is a dilemma many firms have often grappled with, at some point in their mobile journey. In a well-publicized example, the Indian retailer Myntra had tried a bold mobile-app-only strategy but had reverted to both an app and a website presence. They did so after severe push back from their customers for a dual channel presence.[23]

Industry case studies have produced conflicting reports. While Comscore produced a report favoring investments in mobile apps, Morgan

Stanley produced a report favoring investments in mobile Web browsers. The Morgan Stanley report claimed that mobile browser audiences are twice as large as app audiences and have grown 1.2 times faster over the past 3 years and predicted that most publishers will see the majority of their mobile traffic coming from mobile Web browsers rather than from apps.[24] The latest European data from ComScore (July 2016) echo similar findings in a survey of US shoppers by Forrester (August 2015), both of which showed that more users are choosing to make purchases via the mobile Web than via mobile apps.[25] In contrast, a Criteo study from Q4 of 2015 finds that apps account for 54 percent of mobile sales among its clients, ahead of mobile browser 46 percent.[26]

How can the anomalies be explained? It turns out that Criteo's data are based on numbers of transactions rather than on numbers of unique users. This implies that, although more people tend to use the mobile Web to browse, mobile app users tend to buy more. The Criteo study shows that mobile app conversion rates are not just higher than mobile Web, but higher than desktop also.

An academic study by Sinan Aral and Paramveer Dhillon sheds light on this question of whether businesses should invest in a mobile app, in a mobile optimized website, or in both at the same time.[27] It turns out that the impact of a simultaneous presence of mobile app and mobile website has a very nuanced effect on consumer behavior. Using the *New York Times'* digital channels (mobile app and desktop + mobile website) as an example, these researchers showed that mobile app and mobile website act as substitutes. This is due to the fact that people typically consume content via mobile devices in a dynamic environment (e.g., in a subway car or at a bus stop) and so have a limited time budget, which increases the switching cost between the app and browser mobile channels, hence making them substitutes. In contrast mobile app and desktop website are complements. This complementarity arises as people start reading the article on the mobile app (for example, when they're in a busy environment) but finish reading it in more relaxed environment of their home or office on the desktop browser. The complementarity between the channels is significantly less for heavy mobile app users. Tablet users actually end up substituting between the channels.

So let's go back to the original question: What should a firm do? I believe the answer, though a cliché, is "It's complicated." It really depends on your goals.

If measuring the effectiveness of your mobile campaign is the goal, then there are five criteria that should be taken into account when making this decision between an app and the Web: user engagement (higher on mobile apps), attribution ease (easier on mobile apps), relevance targeting (easier and more precise on the mobile Web), creative capabilities (more effective on mobile apps), and ad blockers (a big menace on mobile Web).[28] This framework was first produced by inMobi.

Adoption of your mobile app is, of course, conditional on your app's being used in the first place. It is now a well-established fact that consumers use only a small fraction of apps that they have downloaded. Hence, getting users to discover the app, download it and then consistently use their app is a non-trivial problem for almost any business. Yes, typically only a brand's loyal customers will download their app. But it is important to recognize that these loyal customers are also the biggest spenders. So a lower volume of app use can be compensated by a higher spending amount per app user.

One argument I have also heard in favor of mobile websites is that they show up on Google search. While true, this argument is not going hold water for long. One of the most fascinating advances that will affect the world of mobile apps is deep linking within mobile apps. Deep linking is the idea that a search engine query will show results from within the mobile app for any given business that has the relevant content. Rather than searching first on Google or Bing and then having to manually switch from the mobile browser to the mobile app, deep linking will allow the user to go to specific content or experiences within an app rather than the generic app homepage on the Apple App store or the Google Play store. The large tech companies like Google, Facebook, Apple, and Microsoft are all working on products related to deep linking.

A number of interesting case studies of deep linking are emerging. Will Lindemann describes a number of them.[29] He describes how Yummly increased its user retention by 35 percent by using deep linking to transfer recipe ingredients from Yummly recipes into the Instacart shopping cart and thereby smoothing out the recipe to ingredient purchase experience. Jet, the dynamic ecommerce firm that got acquired by Walmart in August 2016, increased their daily downloads by three-fold and doubled their in-app conversion rates by deep linking from a smart banner to convert a user from the mobile optimized website to the mobile app. Lindemann also mentions how, in order to incentivize users to share video content, Fox Sports uses deep linking to directly link to a specific video in the app when a video is shared. There is no doubt in my mind that new frontiers are going to emerge in mobile marketing as app marketers harness the potential to seamlessly connect traditional marketing, Web content, and the physical world.

So let's get back to the use of mobile apps as a key ingredient in the "Tech mix" force. Can businesses influence consumers to make the extra effort to adopt their app? Does app adoption lead to higher sales as opposed to simply having a mobile optimized website?

A recent study sheds light on these questions. Using a randomized field experiment involving 250,000 customers in the United States, Tianshu Sun, Lanfei Shi, Siva Viswanathan, and Elena Zheleva investigated whether and how a firm can motivate customers to adopt mobile apps and the causal effect of induced mobile app adoptions on customers' purchase behaviors.[30] There are two distinct ways in which firms can incentivize users to download their apps; by providing information about the benefits of the mobile app or by giving explicit monetary incentives. The researchers found that providing information about the benefits of mobile apps and monetary incentives can lead to significant increase in customers' mobile app adoption, with the relative increase being 146 percent and 447 percent, respectively.

Interestingly, this team also found that the true effect of mobile app adoption varies greatly depending on the means by which customers are motivated. Although providing monetary incentives leads to larger increase in mobile app adoption, such induced adoption does not result in more purchases in the long run. In contrast, providing information about the app benefits leads to effective mobile adoption that sustainably increases customers' purchases, and overall profits of the firm.

Even more interestingly, their work showed that app browsing complements desktop browsing. Many users browse on the app but purchase on the desktop. Recall the work that my colleagues and I did with Alibaba, and the importance of cross-device attribution. The finding of this team too, based on data from a leading social commerce platform in the United States, points to the importance of cross-device behavior and attribution.

The similarity in user behavior between United States and China, despite the obvious geographic and cultural differences, is telling. I keep saying that when it comes to mobile, people are fundamentally the same everywhere. Humans we are hard-wired to think and act alike.

In the world of B2B marketing, mobile apps can be used very effectively to provide prospects and customers with usable tools such as tip sheets in order to increase engagement with them. Lee Odden gives us a number of interesting examples in his blog article.[31] He talks about Access, a Fedex mobile magazine app that offers interactive features designed specifically for tablets and Android smartphones, including videos and dynamic slide shows. Power More is a Dell content platform that provides customized content to technology decision makers optimized for mobile devices. 3M's Post-it Plus app for iPhone and iPad enables executives to capture images of their Post-it notes, share and collaborate with other business people.

A recent report published by eMarketer eloquently describes how B2B mobile marketers will use adaptive content (content that is personalized according to the customer's context) in order to provide more flexibility in the omni-channel journey of buyers.[32] From what I have seen, content marketing and email marketing, despite having been around for a while, remain two of the most effective strategies to reach and engage B2B prospects. Given the broad range of devices and screen sizes available in the world of mobile marketing, my prediction is that adaptive content will become a very powerful means for B2B marketers. Mobile will thus play a very important role.

Food for Thought

Are the acquisition costs worth the revenue that Linda will generate, both in the immediate and the longer term, or are some firms essentially gambling on a long shot? Firms can now develop their own robust models to make these kinds of determinations. Attribution is the key to solving Wanamaker's riddle. As Phil Hendrix aptly describes, by establishing cause and effect, attribution answers the following four questions: Which customers are most responsive to particular engagement efforts? How do we design initiatives that maximize outcomes? What are the most cost-effective channels for engaging consumers? What is the value of specific strategies (customers × initiatives × channels) in terms of incremental revenue, customer loyalty, and ultimately customer lifetime value?[33]

In sum, solving Wannamaker's riddle is the holy grail of advertising. If a consumer has been relentlessly bombarded with irrelevant ads or generic offers, it is because brands had very little information about his or her preferences. Consumers can alleviate their pain by building a mutually compatible and endearing relationship with firms such that they can engage them with personalized and constructive services and content. They can help firms help them.

There are no absolutes or guarantees right now, any more than the executives for a major-league baseball team can guarantee a World Series victory because they employ sabermetrics or other "Moneyball" analyses when they assemble their roster. No one wins on paper, and you still have to play the games, as the saying goes. Competition still exists and is never constant, quality will always matter, and the intangibles that help define likeability and loyalty still require further research. But you can make your odds of success much better when you plan an advertising campaign and allocate your budget using the nine forces that have been described in part II.

Takeaways for Firms

Run experiments, analyze your data, and use your domain expertise to extract actionable insights. The good news is that as more and richer data become available, marketers and advertisers can move away from advertising alchemy—that time-worn combination of logic, common sense, experience that allowed them to interpret disparate data—and start measuring effects and differences directly. I recommend firms think about a series of inter-related questions keeping the following things in mind:

- Web and mobile advertising act synergistically and affect each other positively in driving engagement and sales.
- Tablets act as complements for smartphones but as substitutes for personal computers. Invest in device specific advertising accordingly on the basis of where in the path to purchase is the consumer using which device.
- Different ad formats yield different levels of effectiveness depending on the product type. Native advertising has demonstrated huge potential.
- Consider using a combination of a mobile app and a mobile optimized responsive design website. Don't make it an "either or" strategy. For B2B customers, apps can be particularly effective for adaptive content marketing.

III Next-Generation Technology Forces

13 The Growing Intimacy between Us and Our Devices

The possibilities created by mobile technology today are exciting enough. But the future that may emerge by leveraging the nine forces discussed in part II, and new upcoming technologies, is mind blowing.

Before we visualize the future, though, let us take a quick look at the past to get a sense for how fast and how stunning this technological progression has been. In part I, we looked briefly at a device called Simon, a forerunner of the smartphones we use today. It was an important part of a larger progression that has seen us become more intimate with our communications devices.

Three decades ago, we had our communications devices *near* us. Except for the Walkman or the transistor radio, most of our devices were in fixed locations and had prominent on-off switches. We hung up the phone, turned off the TV, and powered down our computer and our game consoles. In the mid 1990s, we started taking our devices *with* us. Phones, computers, televisions (in the guise of DVD players), and gaming devices started to travel with us, making 24/7 connectivity realistic for the first time. We now have the technology *on* us. What's next?

In the next few sections of this chapter, I describe several emerging technologies that soon will be integrated with the mobile ecosystem and will further amplify the power of the nine forces. I foresee the emergence of a few new forces, or at least newer instantiations of some of the existing nine forces. All these technologies will create captive audiences, enhance connectivity, harness more robust data from new sources, and offer better experiences. Business solutions will thrive in the mobile ecosystem, which will remain the hub for many applications and will power amazing business and societal solutions. This list of emerging technologies is by no means meant to be exhaustive. It is only meant to be suggestive.

Some of the biggest advancements will come via wearable technologies that will harness consumers' biometric, neurological, and physiological data, giving birth to a new force that leverages a consumer's physical state of being. Artificial intelligence technologies will make it possible to harness consumers' emotions, and the data generated in the process will give rise to new advertising solutions. Messaging technologies and apps will continue to evolve, and these advanced ecosystems will harness unprecedented amounts of data as consumers spend significant amounts of time on these platforms and share a more intimate side of themselves. The consumer Internet of Things (IoT) will do the same and will harness data from consumers' physical environments, taking efficiency to new levels. Finally, technologies that bridge, connect, and bring together the offline and online worlds will enhance consumers' experiences and create a seamless new hybrid world.

Which countries will lead the way? When it comes to innovation in mobile and related consumer-facing technologies, I have long believed that China and South Korea serve as a crystal ball for what is soon to come in the United States and in the rest of the developed world. For years we have been told that all the meaningful technological innovation happens in Silicon Valley, and that China and other countries follow suit. But in the mobile space, more often than not it is the opposite. China has been at the forefront of innovation and Silicon Valley has often been the follower. One observer, Carmen Chang, a partner at the venture capital firm New Enterprise Associates, echoes my observations well: "China was able to develop a lot of innovative business models, which arose in a different kind of economy. Whether or not we admit it here in Silicon Valley, it's had an impact on us and our thinking."[1] In their book *China's Next Strategic Advantage*, George Yip and Bruce McKern superbly capture how China is establishing itself at the forefront of technological innovation.[2]

14 The Next-Generation Technologies

Wearable Technologies

While mobile devices usher in ubiquitous connectivity, wearable technologies enable seamless connectivity. Consumers no longer need to reach into their pockets to answer a call, change the music, find a shop, or check directions. They can be always on and hands-free at the same time, especially when they can combine wearable technologies with advanced voice-recognition capabilities. But, more importantly, this technology enables a consumer to be more in sync with his or her physical self.

Today consumers are using wearable technologies to record data on their calorie consumption, their heart rates, and their movements and are synchronizing the information with their mobile devices and other connected devices. As these fitness apps learn our habits and precisely document our offline trajectories, they accumulate a treasure trove of data—data that can make message targeting more personalized—and presumably more helpful—than ever before. Many consumers are on the verge of making 24/7 biometric monitoring routine. Once people feel comfortable making those data available for analysis in a larger pool, businesses can understand advertising response rates and effectiveness as a function of consumers' physical condition and state of mind.

With all these advances in data collection, I expect a new force to emerge.

Physiology, biometrics, and neurology are the main ingredients of such a force. The advertisements people watch on TV will be customized on the basis of real-time physiological and emotional data provided by their wearable devices and transmitted through the smartphones to addressable TVs. Conversely, businesses will be able to quantify the physiological and neurological impact of advertising on consumers, adding a new layer of meaning to good ads and bad ads. What an exciting future!

Health and wellness are the most obvious adjacent markets, as this new generation of devices and apps leverages the availability of body placement to measure heart rates and other biometric data. These data will not only track training progress, but also help predict potential health conditions, such as asthma or heart problems, that may otherwise have gone undetected.

Beibei Li, Xitong Guo, Yuanyuan Dang, and I have examined how wearable technologies affect consumers' health care and their wellness. In 2016, we ran several sets of randomized control experiments in which a randomly selected pool of users were given wearable devices and then shared their wearable app data with us.[1] Users in our study were also sent messages from health-care providers reminding them to share their data. There was variation in how the messages were targeted, one group of users receiving generic health-care messages and another group receiving personalized health-care messages based on their individual data. We then matched users' adoption of wearable devices with their health-care and wellness outcomes, both in the short term and in the long term. Short-term health-care data included blood glucose, resting heart rate, and blood pressure. Long-term behavior data included doctor visits and hospital visits.

The study revealed that adoption of wearable technologies leads to changes in lifestyle and to better health-care outcomes (lower levels of blood sugar and blood pressure, and fewer hospital and doctor visits in the long term). Because the participants were selected randomly, there was no concern that we were studying a biased pool of fitness buffs or health-conscious individuals. As one can imagine, this finding is extremely encouraging. If adoption of wearable technologies leads consumers to change their lifestyles and to improve their health care and their wellness, this mobile-induced consumer technology may turn out to be a game changer.

Merely notifying the wearer about a sale on running shoes or giving some new training tips seems far too narrow in this new context. Fortunately, retailers are starting to realize the bigger opportunity and experiment in this space. Ralph Lauren, Tommy Hilfiger, and Under Armour are already beginning to embed chips in apparel to help assess and give advice on how to improve athletic performance, and perhaps to make preliminary medical diagnoses. Running clothes might include chips that advise on a change in gait or breathing. Samsung, Google, OMSignal, Sensoria, Hexo Skin, and other companies are also partnering with brands and investing in this space.[2]

But with every opportunity come challenges. With the smaller screens of wearable technology grabbing more attention, marketers will face a new set of design challenges. If the cognitive effort needed to search through results is greater on a smartphone screen than on a personal computer, it follows that it is even greater when the visual field shrinks to the size of a watch or a display in a pair of eyeglasses. It makes sense that these smaller screens will place a premium on saliency. Marketers will have to tailor programming and messaging to accommodate and take advantage of these new forms of media and opportunities for connectivity. Companies have already begun to resize ads for smaller screens, understanding that the alert on a smart watch or flashing across the field of vision may be the split-second notification the user or wearer grants them. How do you make these messages non-intrusive yet informative when you have so little space and time to work with? How do you maintain a presence and extend the interaction with an app when people quickly scan messages and interact only by tapping or swiping? Advertisers will have to meet this challenge and walk this line as long as the screens we wear are small.

One way to meet the challenge is by fully exploiting apps to create or take over occasions. This keeps a product or a company at the forefront of the user's mind. The more occasions an app provider can win or monopolize, the more influence it can have over the user/wearer. The Sheraton Hotel family's app—accessed either on a smartphone or on a wrist—allows the user/wearer not only to book, confirm, and pay for a stay, but even to unlock the room. The app is also left on and remains alert during the guest's stay. This allows for push notifications of offerings in nearby stores or in a hotel's lobby. The hotel can refresh the daily menu for the restaurant downstairs each morning, and can offer coupons or incentives for future stays. The entire experience becomes more personal, all through an app accessible without even reaching into one's pocket. As the app on a wearable device begins to serve as a butler, the ads themselves will become less of a nuisance and more of a service.

Artificial Intelligence

Remember the chapter on crowdedness and the desire to escape the crowd? An important ingredient of that force is our emotion that propels us to seek that kind of avoidance behavior. While emotion is embedded in crowdedness, in the next few years the situations in which firms can tap into our emotions will also broaden in scope. Hence, "emotion" as a force will also deserve a shoutout in its own right. Can marketers tap into emotions in real time to curate and generate offerings for their consumers? My answer is an emphatic Yes. A fascinating artificial-intelligence-based ecosystem is emerging that will be able to combine real-time biometric data and data on facial expressions with existing behavioral data.

In the future, advertising on mobile devices will be tailored to a user's current emotional and physiological state. This will be done by leveraging biometric data transmitted by wearable devices and consolidating them with emotions extracted from images in real time.

Affectiva, a spinoff of MIT's Media Lab, has created technology to recognize people's emotions by analyzing subtle facial movements. The marketing implications of reading facial expressions and mapping them into emotional states in real time are huge. Wouldn't it be something to be able to customize ads and offers based on consumers' real-time emotions that are captured through the camera in the smartphone or the tablet? I reckon this will become a powerful force that shapes the mobile economy. Affectiva's facial expression technology is already being used for marketing purposes in a mindful meditation app and a live-streaming chat app.[3] In addition, Affectiva's emotional intelligence systems have been adopted by independent video game studios and by brand or advertising teams within large corporations such as Unilever, Kellogg's, Mars, and CBS.[4] Emotient, another startup in this space, has developed software that can recognize emotions from a database of micro-expressions that happen in a fraction of a second. Emotient has partnered with the Honda Motor Company and with Procter & Gamble to gauge people's emotions as they try out products.[5]

The combination of what we stream or watch and how we respond to it can feed into an algorithm that can filter out potential advertisements, letting only the most appropriate and helpful ones through. In essence, our minds and bodies will serve as our own personal ad blocker. In order to do so, those who accept the technology automatically will have to sacrifice some privacy.

This is the case with voice-recognition software that will power such advancements. This is invasive to some, but it reflects that balancing act between using privacy and data as a currency to make our lives easier and more efficient. Devices and apps that respond to voice are becoming more and more common, with Siri and Google Voice on many of our smartphones, and Amazon Echo and Samsung SmartTVs in our homes. Apple is preparing to open up Siri to external app developers, which is clearly going to be a game changer. Virtual assistants like Siri, Google now and Cortana will get trained to pick up contextual clues from our behavior within homes to predict user needs and deliver local information, even before we may actually realize a need. Samsung has gone so far as to collect and transmit unrelated conversations to programming, sending those conversations to partners and companies for purposes of recognition by third parties. As that company states in its Smart TV service policy, "Samsung may collect, and your device may capture voice commands and associated texts so that we can provide you with Voice Recognition features and evaluate and improve the features."[6] Those who engage with these features must be willing to trade some privacy to do so. As voice recognition enters the mainstream and becomes more and more powerful, I fully expect and encourage marketers to be there.

Our predictability can make our lives easier when these devices can proactively manage appointments, create shopping lists, arrange entertainment around moods they anticipate, and even offer expert advice. If that last point sounds far-fetched, witness what happened at Georgia Tech in the winter of 2016. Students were "flabbergasted" to find out that Jill Watson, a very helpful and appreciated teaching assistant for an online course, was not a twenty-something grad student but was, in fact, a robot.[7]

All of this seems like a logical precursor to devices that will be *in us*. This is the technological progression from *by us* to *with us* to *on us* to *in us*.

Instant Messaging and the App Wars

There should be no doubt in anyone's mind that the app economy has had a profound impact on all meaningful aspects of our lives. In chapter 2

I described the typical day in the life of a regular person. Every single aspect of that description is enabled by a mobile app.

Consumers currently have to deal with the problem of too many apps. People use only a small number of apps regularly, and the clear winners are social media and messaging apps. The numbers speak for themselves.[8] Two thirds of digital consumers in China are using the messaging app WeChat for everything from online shopping and VoIP calling to making reservations, paying bills, scheduling appointments, or calling a cab.[9] Hundreds of millions of subscribers around the world have made messaging apps the fastest-growing trend in the mobile ecosystem. Europeans use WhatsApp because text messaging is not free, SnapChat (US) promises privacy. Among the other messaging apps are LINE (Japan and Southeast Asia), Kakao Talk (South Korea), and global contenders such as Tango, Kik, and Viber.[10]

Many messaging platforms appear to be evolving into stand-alone social networks while still retaining the sense of intimate conversation groups. They offer opportunities to provide more personalized offerings, more customized delivery, and more opportunities for efficiency. This makes them very attractive to marketers, who want to tap into these flourishing communities and individuals at the same time.

As these platforms evolve, consumers will spend more and more time in these ecosystems, thereby generating a rich data trail that enables unprecedented levels of personalized advertising. The West is taking a page from the Far East. Facebook now seems to be setting itself up as the West's WeChat.[11] As one observer recently said, "For the Facebook Messenger app, for example, the best way to understand their road map is to look at WeChat."[12] There are other examples too of China's leading the way in innovation. What SnapChat and Kik are doing now with bar-code technology to connect people and share information was pioneered by Alipay and WeChat.

A few years ago, WeChat let third-party retailers and merchants create apps within apps and set up their own individual e-commerce shops. For example, a merchant with an official account on Weixin (the version of WeChat offered in China) can now sell products to customers without having to direct them outside of the app.[13] I expect similar developments in the United States. We are already seeing some of that play out with Facebook's new marketplace. WeChat recently introduced functionality that lets users open apps within WeChat simply by searching or scanning them.

Consumers do not have to download or install these mini-apps separately, and that saves valuable storage space on their smartphones.[14] I expect other Western firms to follow suit.

WhatsApp and Facebook Messenger, two of the biggest messaging apps worldwide, allow brands to chat directly with customers. Both apps are already evolving as data hubs for consumers' interests and habits. The airline KLM and the vodka maker Absolut are already connecting with potential customers about events and encouraging interactions in fun and innovative ways. This gives consumers exclusive access to new product releases and content that have generated unprecedented response. Uber's integration with Facebook messenger to let folks hail cars through the app is another example of this effort.

Everlane, Zulily, and the Hyatt Group have taken mobile messaging to the next level for their consumers by offering customer service through apps.[15] The focus on customer service has also helped the rise of "chatbots." Chatbots—embedded artificial intelligence programs that communicate with users to provide conversational smarts and help with their tasks—make messaging more functional, more personalized, and more entertaining, and they keep consumers engaged longer.[16] Marketers have started to adopt chatbots to better serve customers in a personalized interactive manner. A recent survey reveals that 22 percent of users in financial services, 15 percent of users in health care, and 17 percent of users in retail are engaging with firms through chat apps and chatbots.[17] It seems inevitable that these apps will continue to evolve, increasing the density of functionality and making them platforms for all forms of social commerce instead of mere communication/chat channels. Of course it remains to be seen whether automated chatbots actually increase customer satisfaction. They remind me of automated phone menus, which sometimes are more frustrating than simply requiring a customer to punch in numbers.

In sum, messaging apps will elevate the mobile advertising landscape by cultivating captive audiences and leveraging intimate insights generated on their platforms into highly personalized and tailored experiences.

The Consumer Internet of Things: Smart Homes and Connected Cars

Consumers spend a lot of time in their homes and in their cars. When the homes they live in and the cars they drive get seamlessly connected to their mobile devices and online worlds, new opportunities present themselves. Businesses can get closer to their customers by harnessing the ability of smart homes to solve some common, every day problems. In the process they may also create the gift of time in some cases by making our lives more efficient. As the consumer Internet of Things (IoT) generates data and connects the online and offline worlds, our house serves as a collection and transfer point, an observation platform, and also a device on its own which we can manipulate and manage. By giving people what is truly valuable to them, personalized marketing is going to take a quantum leap thanks to the consumer IoT.

Amazon, Google, Apple, and Microsoft all view the home as the future battleground for a war to push services rather than hardware.[18]

Within a smart home, a consumer can use Amazon Echo to assist with entertaining children, to get step-by-step recipes, or simply to play a song. Apple is also working on a device with a speaker and a microphone that people can use to turn on music, get news headlines, or set a timer.[19] Our homes, robots, and communication systems will learn our habits, preferences, and tendencies over time. This vast amount of data will reveal even more about consumers' predictability, no matter how spontaneous they may want to be. They can also help us simplify and take care of occasions that may seem special on the surface but have a lot of routine components. Imagine that one of these in-home devices has fielded enough requests from one of your children to make inferences about what would make an ideal birthday gift. Full integration with your mobile data would allow the device to explore online deals, make price comparisons, prepare a short list of suggestions for you, and even figure out, on the basis of family calendars, whether the gift should be shipped to the recipient's home, to your office, or to another address. When you approve, it can order it, pay for it, and arrange the shipping. Does this diminish the gift's value when a machine chooses it instead of you? That is a question I cannot answer, except to say that some people will appreciate the efficiency and the opportunity to spend that time saved with their child in a high-quality fashion whereas others will prefer to buy a gift the old-fashioned way.

Will consumers go for it? Will the personalization enabled by smart homes trump privacy concerns? I believe so. I admit I don't necessary know exactly what that personalization might look like, but if consumers' history is any indication it will happen. Donna Hoffman and Tom Novak describe it well in an excellent article on the future of the smart home and the consumer IoT[20]:

> We believe identities for the smart home have the potential to emerge that are likely to outweigh privacy concerns. The shape of these identities may give some insight into what features consumers value enough to trade off some aspects of their privacy for. That is, we believe that "something more" will emerge that will be more valuable to people in their homes than their concerns about privacy. While it is difficult to speculate, we think these identities will possibly include personalization efforts. We believe that as smart home assemblages, largely through habitual repetition, come to know more and more about the inhabitants in these homes, the potential benefits from personalization could very likely trump privacy.

Like many other pioneering technologies, the consumer IoT is almost ready to tackle inefficiency in daily mundane tasks such as commuting. Commuting in an automobile can be a waste of time. Even in a modern, Wi-Fi-enabled vehicle with a smartphone nearby, the best drivers remain digitally impaired. Just how much time do these commutes cost us? The management consulting firm A. T. Kearney estimates that self-driving cars will free up 1.9 trillion minutes of idle time in the year 2030. When that happens, it will provide marketers another set of opportunities to vie for consumers' attention while creating additional hours in the day for every consumer. Of course that is not the only upside of self-driving cars; a reduction in the number of accidents is generally touted as the greatest benefit.

Connecting our cars to our personal mobile and wearable ecosystem can improve our daily lives. Ford is considering taking advantage of the data collected through biometrics to make driving much safer.[21] Data on sleep patterns, alcohol purchases, and other data can be analyzed and interpreted to make suggestions aimed at ensuring better experiences on the road. The connected car will also communicate through the mobile device. Ford is considering offering a connection to Amazon Echo, which would allow for customized weather updates, shopping list additions, and temperature control,[22] and is working on integrating Amazon Alexa into its cars. Volkswagen recently announced a partnership with the Korean electronics manufacturer LG to integrate smart homes with smart cars, thereby allowing drivers to monitor their homes from their cars. For example, as a consumer leaves the office to begin her homeward commute, she can turn on her home's lights and get the home preheated to her desired temperature.[23]

An article in the magazine *Mobile Marketing* sums it up well:

In the era of big data, programmatic targeting and location-based ad campaigns, being able to connect to motorists and passengers in real-time and harvest information like what music they're listening to and how often they're checking maps can provide marketers with additional tools to make advertising more accurate and effective.[24]

Because of their ability to solve real pain points for consumers in their everyday lives, smart homes and smart cars will create new sources of data from our physical environments. By having the ability to have customized real-time solutions, the IoT will create new opportunities for monetization in the mobile economy.

Technologies That Bring Our Offline and Online Worlds Together

In the coming years a number of new technologies will focus on closing the gap between our offline and online worlds and on creating an omnichannel view of each consumer. Two of these technologies are smart wallets and virtual reality/augmented reality (VR/AR).

From smartphones to smart wallets

Paying for purchases is becoming just as seamless as the rest of our lives. Thanks to Apple Pay, Android Pay, and Square, mobile payments systems are now poised to cause a huge disruption in the "fintech" space. As wearable devices make it possible to pay without removing a wallet or accessing a phone, the propensity to use this technology grows even stronger. Nearly 80 percent of Apple Watch users use Apple Pay to pay for both online and in-person purchases.[25] Credit card and debit card information is stored in a Wallet, a function of the Passbook system that allows for everything from boarding passes to concert tickets purchased on Apple Pay. This makes purchases even easier to make from mobile or wrist without removing or even carrying the physical card. Android is following suit with the Android Pay system. Security is provided through PINs and other codes not stored on the device. Both major mobile technologies are developing ways for customers to walk through a physical store, select an item, tap a phone or a wrist app to scan a bar code, and then make a purchase without even thinking of waiting in line. Mobile pay offers a gold mine of data on consumers' behavior and unprecedented direct access to shoppers. This makes the path from click to conversion seamless as well, and can allow marketers to integrate more offline purchases into their data sets.

From the strap on your wrist to a strap around your head

The premise is simple. Businesses and marketers need to be wherever consumers will be. As more and more consumers adopt an awkward looking strap around their heads to enjoy rich, immersive experiences, content creators and advertisers will follow them and develop experiences that leverage mobile devices and such items as Google Cardboard and Samsung Gear virtual reality apps. Today the VR experience on mobile is not as rich as on desktops but this is going to change in the coming years as VR developers and manufacturers focus on mobile. Mobile VR (a VR headset utilizing a mobile device to run the virtual reality experience) will become more prevalent. Facebook Oculus and Samsung Gear VR are only the tip of a massive iceberg.

VR/ AR has far-reaching implications aside from transforming the world of content. One of the biggest challenges for businesses in some categories today remains their inability to showcase their goods or services in a way that a physical store can. This challenge only gets bigger as our devices and smartphones get smaller and as retailers cut back on new stores, close existing stores and invest in e-commerce. Any industry that benefits from experiential marketing—real estate, home improvement, or travel, for example—will adopt VR/AR. In essence VR/AR will bring the offline to online.

Conversely it also can bring the online to the offline and solve one of the greatest challenges that businesses face today: how to increase foot traffic in physical stores. The initial popularity and subsequent crash of Pokémon Go notwithstanding, there are going to be ample opportunities for firms to use VR/AR on smartphones to integrate our physical worlds with the digital world and disrupt marketing in a non-trivial way. We have seen that wildly popular mobile games such as Pokémon Go can drive insane amounts of store traffic and sales in small, local businesses.[26] This will promote the possibility of sponsored locations on a cost-per-visit basis, similar to Google's cost per click payment model.[27] Sponsored location means that offline stores pay to become Pokémon stops within the virtual game board. As Niantic's CEO John Hanke says, the concept is rooted in the premise that being a sponsored location is "an inducement that drives foot traffic."

In summary, there is a lot to explore with the technology and capabilities at hand today in the mobile world. At the same time, there is a lot in the future to be excited about as well. Within a few years, the mobile ecosystem and related technologies will transform our lives beyond recognition and usher in a new age. It will without doubt be recognized as one of the hallmark advancements that society has seen in the 21st century.

Epilogue

Do you remember Tom and Rachel from chapter 9? Tom and Rachel were walking their dog and were going to Starbucks rather than to Chipotle. Now imagine that, a few years later, Tom enters their apartment after a long run on a weekend. Rachel is watching the fiftieth-anniversary reissue of the movie *Marathon Man,* and she and Tom laugh as they watch Babe return to his apartment after a long run. Babe grabs a clipboard and pen and writes down some notes from his run—presumably the time and the distance.

Tom, of course, didn't need to do that. His implant had recorded everything about his run in minute detail, from time and distance to various biometrics (heart rate, blood sugar) to the weather conditions (humidity, headwind, tailwind) to course conditions. You get the point. The device compared today's run with his own personal history, with other runners who ran the same course, and with other 40-year-olds nationwide. It recommended a nutritionally balanced snack and dinner for him, keeping in mind that he will have a short night's sleep because he has an early morning flight the next day from Nashville to Los Angeles. Two items for his getaway breakfast were missing from their refrigerator, so some device in their home had the missing items ordered and delivered. That may have been done by the refrigerator itself, but more likely it was done by Tom's Pebble Smart Watch, which can integrate with Amazon's Alexa service and seamlessly place orders.[1] Tom and Rachel never even knew that the two breakfast items were missing until the delivery arrived.

Tom and Rachel, like billions of other people around the world, have become nodes in a gigantic seamless network, directly or indirectly linked to things and to people who have a strong influence on their lives, even though they will probably never see those things or those people. The debate will rage, as it has since the Industrial Revolution began in earnest, about whether the technological progress that led to this outcome is a "good thing" or a "bad thing."

Regardless of which side you are on, this technological progression will continue. It is unstoppable. In 2016 we are still at the rudimentary stages of understanding where the apps on our smartphone can lead us, and what tasks we can delegate to that concierge they reside on. What will our world look like when these apps are accessible on our wrists, in our glasses, and through our homes and cars, through devices we haven't thought of yet, or even within our own bodies? I won't indulge in the details of what implants or our world in 2026 might really look like. That task is best left to talented science fiction writers. But I can leave you with a few closing thoughts about the impact of mobile technology on society at large.

Can Your Favorite Brands Keep Your Secrets?

The example I outlined above is a microcosm of what our worlds will look like in a few years. There are going to be many other opportunities. But there will be challenges as well. Many of the forces I described in this book can give rise to privacy and security concerns among some of you readers and may take the form of a new societal issue, so it bears mentioning the differences between personal data collected to improve and contribute to your life versus malicious security breaches perpetrated by hackers. It is imperative that we distinguish between concerns stemming from data security and data privacy. There are some subtle but important differences.

Many of us have heard about breaches in data security at major businesses—retailers, health insurance companies, financial institutions, etc. involving hackers getting access to our transaction data or our personal data. We assume that the worst has happened. It does not help that certain sections of the media sometimes sensationalize such incidents more than they should, probably to drive additional eyeballs. Consumers are often misled to believe that successful data breaches are regular occurrences, but in reality they are relatively rare.

When consumers entrust a business with their sensitive private information, the business must have an effective security policy to protect the data. Consumers' concerns about data security will be alleviated through infrastructure investments that prevent security breaches and hacks. This issue needs to be addressed head on. Businesses are also actively pursuing perpetrators of security crimes and identity theft, improving security measures to prevent or deter hackers, and protecting our data in many other ways.

Consumers' concerns about data privacy, on the other hand, should be about how firms are harnessing their data to enhance consumers' lives. Are they using data judiciously to add value and make relevant offers?

What may seem astonishing to many people is that most businesses have very little incentive to abuse our data and violate our privacy intentionally. For all the reasons I have outlined in this book, having an intimate connection with consumers can give a company a huge competitive advantage, and therefore the company has very little reason to intentionally jeopardize that. All a business intends to do is to create the right offer at the right place to the right person. At the same time, businesses also need to recognize the tradeoffs between offering privacy and a concierge service. They need to be cognizant of the "slippery slope" between a butler and a stalker. They need to understand where to draw the line and ensure that they don't cross it. They need to do everything possible to ensure that consumers are protected.

Apple, which has never been as aggressive as Google, Facebook, or Amazon about collecting data from users, is reputedly going to introduce solutions that will make it difficult for anyone (including the federal government) to access our data.[2] Apple also criticizes other tech companies for collecting personal information that is used to target advertising. But that policy has hindered Apple's ability to develop and improve services for users.

Apple is now tapping new technology to garner insights into users' behavior from keyboards, from Spotlight, and from Notes. Referred to as "differential privacy," the technology will be included in a forthcoming update to iOS for iPhone and iPad. It is based on a tool in statistics that uses hashing and sub-sampling to let its data engineers spot patterns on how a group of users is using their devices while making it very difficult for anyone to link accurate data back to an individual user. Apple isn't the first company to experiment with differential privacy. Google has been using the technology for 2 years to analyze some data from its Chrome browser.[3]

Apple's recent acquisitions also signal its commitment to protecting user data privacy. It acquired Perceptio, a startup that makes it feasible to power its AI assistants without having to mine a user's personal data. Unlike many firms, Apple doesn't collect data on its users. If you have wondered why Google Now is so effective, part of it is because it mines data from consumers' search histories, Gmail, calendars, etc. to predict consumers' intentions.[4] The Perceptio acquisition suggests that Apple wants to balance a user's privacy while still providing the user with an excellent AI assistant.

Other firms also maintain a higher standard than established legal requirements. For example, even though Google assisted with 78 percent of law enforcement data requests in 2015, it maintains a hard line toward data sharing and won't disclose users' location history data without a legal warrant.[5] Facebook makes it abundantly clear to its users that if they don't want Facebook to use data from other websites and apps to show more relevant ads, they won't.[6] Users can opt out of this type of ad targeting in their Web browser using the industry-standard Digital Advertising Alliance opt out,[7] and on their mobile devices using the controls that iOS and Android provide.

Edward Snowden made the headlines when he exposed government surveillance measures. Wikileaks enlightened us about similar matters. Since then many consumers have developed an intrinsic (and often unwarranted) fear and knee jerk reaction to what organizations want to do with our data. It is imperative that we understand the huge difference between data in corporate organizations (accrued from our logging in on different digital platforms, sharing personal preferences by engaging with ads, or sharing information by keeping the location button on in the mobile phone) to be used for marketing purposes versus data collected in government organizations to be used for surveillance purposes. We need to understand this difference, and also embrace this distinction between industry's use of data and government's use of our data. In May 2016, the Fourth Circuit Court of Appeals ruled the various location histories assembled through our phones can be requested by law enforcement without requiring a warrant.[8] Now that should worry people more than businesses sending them less relevant offers.

What's at Stake with Mobile Technology?

Despite all the debate around privacy and data, we cannot turn away from mobile technology. As we look more closely at the power and impact of this technology, we see that it delivers not just economic impact, but increasingly has demonstrated political and social impact. While this book has been about business and commerce, I feel compelled to point out that in recent years, smartphone technology has demonstrated several noncommercial opportunities. It has shown what connectivity and data sharing can do to influence actions and reactions. Mobile technologies have shown how it can save consumers a few dollars and sometimes even their lives. It is this holistic impact of mobile technology that makes me so passionate about this topic.

Today there are so many examples of societal impact being brought by mobile technology that they deserve an exclusive book of their own. Yet no discussion of mobile technology is complete without acknowledging the social transformation it has brought about. The examples below are a very tiny sub-set.

The ability to exchange information rapidly in real time across a large, ad hoc network helped trigger the Arab Spring. Locations and check-ins have proved to be so helpful after natural and man-made disasters that Facebook has now implemented a safety feature that allows users to let others know they are safe without the need for triangulated cell towers or for bandwidth that might be at a premium during an emergency. The number of people in a community affected in a disaster, which can use cellular data before they might be able to make a call is substantial. One example is the major earthquake in Haiti. People who had Blackberries could get emails out long before calls were available. Similarly, after the terrorist attacks in New York City on September 11, 2001, email and other means of data communication were the best options. Towers were not disabled, but phone lines on the towers were jammed. When messages are short, such as a quick ping on the Facebook safety feature, transmission is faster and takes up less bandwidth. Data programs will often automatically reattempt to send the message if the first attempt cannot get through.[9]

Mobile applications are also making a direct difference. The Philippines are a chain of islands and lowlands along a fault line, hit frequently with earthquakes and typhoons. Additionally, many of the residencies are considered rural, with limited access to government services or emergency care. There is no government sponsored 911 emergency response or dispatch in the country because of the geography and demographics. Pilipinas 911, a privately owned emergency dispatch company in the Philippines, introduced a dispatch and training model to combat this problem. Because most of the country has access to mobile, those who register for the service are able to call for assistance in case of an emergency and have local ambulances respond.[10] The founders are now enhancing the service even further, to cover household emergencies, such as a stroke or a fall, or urgent responses needed in a national emergency. The app is designed not only to call emergency services in a country where such services don't generally exist, but also to pinpoint a last known location if activated at the beginning of a disaster. An app that uses time, context, location, trajectory, and several other forces to cut search and recovery time could mean life or death in a country such as the Philippines. When a system can locate and dispatch the closest responder, it can make incidents such as Typhoon Haiyan, which is estimated to have killed more than 10,000 Filipinos, less threatening and less deadly. The penetration and ubiquity of smartphones now enables other countries with heavy rural communities and frequent major disasters to implement the system.

Emergency and relief are only one way we might use the concepts from marketing and advertising to effect social change. By 2016, the smartphone has already established itself as an indispensable device for 4.61 billion people on the planet and it will grow to 5.07 billion by 2019.[11] It has penetrated some of the world's most remote and underdeveloped regions. This rapid growth has allowed populations with limited access to information or goods, such as those in rural India or Indonesia, to communicate and trade as part of the global market. Mobile technology can provide fishermen in India real-time information on fish prices and enable villagers in South Africa to make low-cost mobile payments.

In India, mobile allows people to have Internet access without a home computer. While there is still a lot of untapped potential, mobile banking and some online retailers are expanding their reach into the rural Indian market, paving the way for those with limited access to have more economic opportunity.[12] Two researchers from Harvard Business School ran a fascinating a study on the impact of providing toll-free access to AO, a mobile phone-based technology that allows farmers to receive timely agricultural information from expert agronomists and their peers.[13] This mobile access significantly changed farmers' sources of information for sowing and agricultural input-related decisions such as recommended seed varieties, fertilizers, pesticides and irrigation practices. In particular, farmers relied less on commission-motivated agricultural input dealers for pesticide advice, and less on their experience, when making fertilizer-related decisions. The authors found evidence of non-trivial improvements in yields as well: treatment effects up to 8.6 percent higher for cotton, and 28.0 percent higher for cumin. The authors estimated that each dollar spent on the service yields a $10 private return. What an amazing study!

I hope these examples give you a glimpse of how mobile technology can also enable health-care agencies and non-profit service agencies to reach populations without access to home computers, and provide them with better support.

How to Prepare Yourself for the Mobile Age

Throughout this book, I have talked about how the mobile economy can have a profound economic impact by leveraging the key nine forces. To really harness the potential of these nine forces, businesses and individuals need to pay increasing attention to the skills they look to develop.

In his 2002 book *The Future of Work*, former Secretary of Labor Robert Reich described two broad categories of employees who will be in very high demand if technology continued to develop on a rapid upward path. He labeled them as the "geeks and shrinks." According to Reich, "geeks" are the ones who are inventive and are able to take whatever product or service they are working on and make them better, faster or cheaper.[14] On the other hand, "shrinks" are the ones who have insight into what people might want, even though they don't even know their wants, probably because there's no product or service to test their wants.

I firmly believe that the mobile-economy-driven world of the future will need a lot more geeks and a lot more shrinks.[15] To prepare them will require that academic institutions refocus their education systems and their pedagogy. To absorb them into the workforce, companies will have to change their hiring and promotion practices, compensation structures, intra-organizational culture, and make additional investments.

The geeks will certainly have plenty of work to do and plenty of significant problems to solve in terms of database design, data collection, and data analysis. From a data perspective, this rapid technological progress was neither seamless nor victimless, and it has left some tricky engineering challenges in its wake. Someone will have to solve them. Creating self-driving cars, smarter houses, and other devices and bringing them to the mass market will require breakthroughs in sensor technology, software, artificial intelligence, augmented reality, robotics, and a number of other areas. The opportunities are enormous, but the bar is high. In fact, this prompted Vinton Cerf, the "father of the Internet," to warn of a Digital Dark Age if we don't figure out a way to preserve our data and pictures and other memories as technology advances.[16]

Data—essential to powering the nine forces discussed in this book—require a lot of tender, loving, care. Cleaning, managing, and analyzing data are complex jobs. Companies that hope to capitalize on the insights presented in this book—and more new insights, which will surely follow—will need teams of business analysts and data scientists with the skills to build models, analyze patterns, and perform other advanced statistical and predictive analyses. Companies will be inundated with data as they track all the forces required for a successful mobile marketing campaign. They will need intricate modeling systems to better understand and use the information. Consolidation of data is essential.

This new age will continue to require more and more quantitatively trained business executives and managers with different skills than what the traditional economy has needed. With an estimated 4 million data-related jobs available and only about 30 percent of those jobs filled,[17] data-driven managers are already becoming "hot hires" for companies outside of technology. Companies are no longer looking only for managers with an eye for competitive strategy or market experience. They will increasingly need someone who can take the huge swaths of data they are collecting and turn it into useful ways to target, engage, and win over customers.

This is where the shrinks come in. These analyses require interpretation and translation into meaningful messages for cross-functional business teams and consumers. Businesses will continue to need lots of storytellers. The importance of story-telling should not be underestimated at all. Marketers who can combine quantitative literacy with communications skills and a background in consumers' psychology will have an advantage. They will take on the creative challenges that still remain once the data have been analyzed.

Not everyone obviously will be a geek or a shrink and contribute directly to the mobile economy. Yet most people will live in them and benefit from them, in terms of greater opportunities for flexible work arrangements, global communication and interaction, and adding value outside the confines of a traditional office, factory, store, or shop.

Mobile has and will continue to have an important role in the workplace, albeit in different ways. Despite the prominence of desktop in the workplace, mobile will increasingly blur the lines between work and personal time. Thirty-five percent of US employees use smartphones for work purposes outside of their regular work hours.[18] Not everyone is happy about it because spending an additional ten hours per week on work does not necessarily improve the quality of consumers' lives. Vincent Brissot, a vice president of Worldwide Channel Marketing and Operations at Hewlett-Packard, made a poignant remark on this subject: "The statement 'I am off to work' is not true anymore. Mobility radically changes the true nature of work. Work is not where you go anymore. Work is what you do." The debate is ongoing if the benefits of mobile technology in the workplace outweigh the costs it imposes on our lives.

A natural extension of this usage trend is that in order to increase engagement, firms will support adaptive content and other more advanced tactics. And in order to support such personalized content and omni-channel experience, mobile marketers will turn to programmatic marketing (marketing that is automated through machine learning algorithms) even in the B2B world.[19]

In Conclusion

More than 50 years ago, Marshall McLuhan coined the term "global village" to describe a world completely interconnected and interdependent thanks to telecommunications.

That world is here.

When the world has more smartphone subscriptions than it has people with true access for all including the elderly and the poor, we can all reach one another without the friction and frustration that has delayed and distressed human communication since it began. That sounds like a bold declaration, but I hope you have seen enough proof in this book to appreciate the economic opportunities that mobile technology has created. Prospering in the global village will require some effort, however.

Through all these exciting changes, I anticipate that the four behavioral contradictions will persist. The future of mobile advertising depends on a bargain that consumers and firms need to strike with each other. Both sides will have to make some investments and offer some trust for this give-and-take relationship to prosper. Consumers will need to find better ways to strike a personal balance between their lives and mobile technology. They will need to make the choice about how much they let technology intrude and inform their lives. They will hold the key to how open or private they want to be with their data. Most certainly, the onus will be on them to find a healthy balance and keep an open mind as they make those choices.

But this does not relieve businesses of their responsibilities. They need to pay attention and take their roles very seriously in this ecosystem. Consumers are more and more willing to share their data as currency, as long as they receive in return something of real value, such as more entertainment, more efficiency, or savings of time and money. Firms need to uphold their end of the deal. They need to surprise and impress consumers while helping them with their needs the way a butler or a concierge would. More importantly, they need to take data security and privacy issues seriously.

Rarely do we enter an economic era when the upside is so life changing, positive for consumers and society, and also very lucrative for the firms who live up to their end of the bargain. I am confident that all sides will learn the value in honoring the commitment and jointly manage the forces that drive the mobile economy.

Notes

Introduction

1. http://www.smartinsights.com/mobile-marketing/mobile-advertising/7-effective-mobile-marketing-campaigns/

2. https://www.gsmaintelligence.com/research/?file=97928efe09cdba2864cdcf1ad1a2f58c&download

3. https://apac.thinkwithgoogle.com/articles/the-future-of-shopping.html

4. https://www.facebook.com/help/585318558251813

5. http://www.imediaconnection.com/article/230007/160419-ben-plomion-everything-oldis-new-again

6. http://www.smartinsights.com/mobile-marketing/mobile-advertising/7-effective-mobile-marketing-campaigns/

7. http://www.pwc.com/us/en/industry/entertainment-media/publications/consumer-intelligence-series/consumer-privacy.html

8. http://www.pwc.com/us/en/industry/entertainment-media/publications/consumer-intelligence-series/mobile-advertising.html

9. http://www.millennialmedia.com/mobile-insights/blog/whats-my-worth-how-ads-appeal-to-consumers

10. R. Sloan and R. Warner, "Beyond notice and choice: Privacy, norms, and consent," working paper, Social Sciences Research Network, 2013.

11. http://mentalfloss.com/article/54133/6-ways-cell-phones-are-changing-world-beyond-ways-youre-probablythinking.

Chapter 1

1. J. Bezerra et al., "The mobile revolution: How mobile technologies drive a trillion-dollar impact," BCG Perspectives, 2015. https://www.bcgperspectives.com/content/articles/telecommunications_technology_business_transformation_mobile_revolution/ .

2. Ericsson mobility report 2016.

3. Bezerra et al., "The mobile revolution."

4. Ibid.

5. Ibid.

6. http://www.computerworld.com/article/2520954/mobile-wireless/google-ceo-preaches--mobile-first-.html

7. https://techcrunch.com/2015/09/10/u-s-consumers-now-spend-more-time-in-apps-than-watching-tv/

8. https://yourstory.com/2015/06/location-based-marketing/

9. http://www.adweek.com/news/technology/here-everything-you-need-know-about-current-state-mobile-ad-fraud-168542

10. http://www.merriam-webster.com/dictionary/smartphone

11. http://www.bloomberg.com/news/articles/2012-06-29/before-iphone-and-android-came-simon-the-first-smartphone

12. http://fortune.com/2016/01/14/apple-iphone-q1-2016/

13. ProximityMarketinginRetailQ12016Report..

14. http://unacast.com/shhh-plan-tell-everyone/

15. Location Data Privacy: Guidelines, Assessment and Recommendations. The Location Forum, 2013.

16. Ibid.

17. Ibid.

Chapter 2

1. http://www.forbes.com/sites/neilhowe/2015/07/15/why-millennials-are-texting-more-and-talking-less/#4b9815c95576

2. http://www.payscale.com/career-news/2015/01/smartphones-in-the-workplace-productivity-tool-or-time-suck-

3. Some early adopters of LBS include the app Shopkick (founded in 2010), a coupon and deal push mechanism. Retailers such as Macy's, Target, J. C. Penney, Crate & Barrel, and Best Buy could send deals as long as customers had the Shopkick application open on their smartphones and were present in the store.

4. http://www.emarketer.com/Article/Most-Smartphone-Owners-Use-Location-Based-Services/1013863

5. http://www.govtech.com/dc/articles/How-Can-Location-Based-Services-Overhaul-Local-Government.html

6. http://www.govtech.com/dc/articles/How-Can-Location-Based-Services-Overhaul-Local-Government.html

7. http://www.govtech.com/dc/articles/How-Can-Location-Based-Services-Overhaul-Local-Government.html

8. A. Pentland, *Social Physics: How Good Ideas Spread—The Lessons From a New Science* (Penguin, 2014).

9. http://www.economist.com/news/united-states/21615622-junk-science-putting-innocent-people-jail-two-towers

Chapter 3

1. D. Read and G. Lowenstein, "Diversification bias: Explaining the discrepancy in variety seeking between combined and separated choices," *Journal of Experimental Psychology* 1 (1), 1995: 34–49.

2. G. Becker and K. Murphy, "A simple theory of advertising as a good or bad," *Quarterly Journal of Economics* 108 (4), 1993: 941–964.

3. Millennial Media, What's My Worth, 2015. http://www.millennialmedia.com/mobile-insights/industry-research/whats-my-worth-report-2015

4. A. LePage, "Is the medium the message?" *The Programmatic Mind* (http://buyercloud.rubiconproject.com), issue 10, March 4, 2016.

5. What I mean by "mobile Web" (and what is generally meant in the industry) is Internet access via a browser on mobile devices (as opposed to on desktops/laptops).

6. http://fortune.com/2015/09/18/ad-block-ethics/

7. http://www.nytimes.com/2016/08/10/technology/facebook-ad-blockers.html

8. G. Raifman, "Why ad blockers are good for the ad industry," *The Programmatic Mind* (http://buyercloud.rubiconproject.com), issue 10, March 4, 2016.

9. Millennial Media, What's My Worth, 2015.

10. https://www.linkdex.com/en-us/inked/mobile-marketing-trends-2016/

11. http://www.nytimes.com/2016/08/10/technology/facebook-ad-blockers.html

12. B. Schwartz, *The Paradox of Choice: Why More Is Less* (Harper Perennial, 2005).

13. A. Ghose, P. Ipeirotis, and B. Li, "Examining the impact of ranking and consumer behavior on search engine revenue," *Management Science* 60 (7), 2014: 1632–1654.

14. S. Iyengar and M. Lepper, "When choice is demotivating: Can one desire too much of a good thing?" *Journal of Personality and Social Psychology* 79 (December 2000), 995–1006.

15. I. Simonson, "The effect of product assortment on buyer preferences," *Journal of Retailing* 75 (autumn), 1999: 347–370.

16. S. Peters, "Believe it or not, millennials do care about privacy, security." http://www.darkreading.com/endpoint/believe-it-or-not-millennials-do-care-about-privacy-security/d/d-id/1322622

17. http://www.adweek.com/news/technology/digital-savvy-millennials-will-sacrifice-privacy-personalization-says-leo-burnett-exec-169869

18. http://www.tsys.com/ngenuity-journal/millennials-and-privacy-more-open-to-sharing-but-also-more-informed.cfm

19. https://www.research-live.com/article/news/millennials-the-most-privacyconscious-generation-says-study/id/4011682

20. http://www.theatlantic.com/technology/archive/2016/04/would-you-let-companies-monitor-you-for-money/476298/.

21. http://www.theatlantic.com/technology/archive/2016/04/would-you-let-companies-monitor-you-for-money/476298/.

Introduction to Part II

1. http://www.smartinsights.com/mobile-marketing/mobile-advertising/7-effective-mobile-marketing-campaigns/

2. http://www.pwc.com/us/en/industry/entertainment-media/publications/consumer-intelligence-series/consumer-privacy.html

3. http://www.pwc.com/us/en/industry/entertainment-media/publications/consumer-intelligence-series/mobile-advertising.html

4. http://www.millennialmedia.com/mobile-insights/blog/whats-my-worth-how-ads-appeal-to-consumers

5. http://info.localytics.com/blog/5-real-app-examples-of-killer-geofencing-push-notifications

6. http://www.nytimes.com/2012/02/19/magazine/shopping-habits.html?pagewanted=6

7. http://mobilefomo.com/2015/04/types-context-enrich-mobile-marketing/

8. http://www.techrepublic.com/article/mobile-apps-need-context-to-hit-the-right-targets/

9. http://mobilefomo.com/2015/04/types-context--enrich-mobile-marketing/

Chapter 4

1. http://www.cntraveler.com/stories/2012-10-12/cruises-restaurant-quality-food-explained

2. http://www.wsj.com/articles/SB1000142405274870448650457509742329650 6784

3. http://touristmeetstraveler.com/2014/behind-buffet-cruise-ship-food-kitchens/

4. http://www.thedrinksreport.com/news/2013/15201-special-feature-serving-the-cruise-industry.html

5. http://www.foxnews.com/travel/2016/02/26/cruise-ship-food-10680-hot-dogs-just-tip-iceberg.html

6. http://streetfightmag.com/2016/04/07/the-context-for-contextual-marketing-is-changing/

7. http://about.mapmyfitness.com/2013/11/underarmour/

8. http://streetfightmag.com/2016/04/07/the-context-for-contextual-marketing-is-changing/

9. A. Sadilek and J. Krum, Far Out: Predicting Long-Term Human Mobility. Association for the Advancement of Artificial Intelligence, 2012. http://www.aaai.org/ocs/index.php/AAAI/AAAI12/paper/view/4845

10. A. Ghose, B. Li, and S. Liu, "Mobile advertising and realtime social dynamics. A randomized field experiment," working paper, New York University, 2016.

11. http://streetfightmag.com/2016/04/07/the-context-for-contextual-marketing-is-changing/

12. P. Adamopoulos, A. Ghose. V. Todri, and A. Tuzhilin, "The business value of recommendations in a mobile application: Combining deep learning with econometrics," working paper, New York University, 2016.

13. https://www.bostonglobe.com/arts/2016/06/05/why-our-location-histories-are-glimpse-into-future/OqFyyVTy1sWWBJRDguvX5K/story.html#comments

14. http://mobilefomo.com/2015/04/types-context-enrich-mobile-marketing/

15. M. Brodeur, "Why our location histories are a glimpse into the future," *Boston Globe*, June 6, 2016. Google created the term "micro-moments" in 2015 to dramatize the impact of mobile on our search behaviors. See http://streetfightmag.com/2016/04/07/the-context-for-contextual-marketing-is-changing/

Chapter 5

1. https://www.engadget.com/2015/05/22/philips-led-vlc-navigation/

2. P. Zubcsek, Z. Katona, and M. Sarvary, "Predicting mobile advertising response using consumer co-location networks," working paper, Social Science Research Network, 2016.

3. https://www.intersec.com/location-based-advertising-lba-market-to-hit-15m-by-2018

4. https://www.b2bmarketing.net/en/resources/blog/why-sms-your-most-powerful-b2b-marketing-tool

5. https://www.b2bmarketing.net/en/resources/blog/why-sms-your-most-powerful-b2b-marketing-tool

6. http://www.mmaglobal.com/casestudies/avenue%E2%80%99s-rmm-campaign-yields-roi-over-6600

7. http://streetfightmag.com/2015/12/22/10-top-location-based-marketing-campaigns-of-2015/

8. Skyrocket Conversion Rates in 2016 and Beyond with Context. Skyhook Report, 2016. http://info.skyhookwireless.com/hubfs/eBook_Skyrocket_Conversion_Rates_in_2016_and_Beyond_with_Context.pdf

9. http://www.wsj.com/articles/SB1000142412788732377204578189391813881534

10. S. Banerjee and R. Dholakia, "Mobile advertising: Does location-based advertising work?" *International Journal of Mobile Marketing* 3 (2), 2008: 68–74.

11. S. Spiekermann, M. Rothensee, and M. Klafft, "Street marketing: How proximity and context drive coupon redemption," *Journal of Consumer Marketing* 28 (4), 2011: 280–289.

12. J. Lindqvist, J. Cranshaw, J. Wiese, J. Hong, and J. Zimmerman, "I'm the mayor of my house: Examining why people use Foursquare—a social-driven location sharing application," presented at CHI 2011, Vancouver.

13. A. Ghose, A. Goldfarb, and S. P. Han, "How is the mobile Internet different? Search costs and local activities," *Information Systems Research* 24 (3), 2013: 613–631.

14. W. Wang, "QR codes make location matter even more: The mere exposure effect of QR codes," working paper, Hong Kong University of Science and Technology, 2016.

15. http://www.xad.com/press-releases/geo-conquesting-is-the-new-craze-in-mobile-advertising-according-to-the-xad-q2-2013-mobile-location-insights-report/

16. Mobile Retargeting, Optimization & Hitting the ROI Bullseye. YP Marketing Solutions. http://national.yp.com/downloads/MMS_Chicago.pdf

17. http://marketingland.com/using-geo-location-turn-mobile-traffic-line-gold-80820

18. http://www.geomarketing.com/what-is-geo-conquesting

19. http://medicomhealth.com/geotargeting-for-healthcare-marketing/

20. http://www.xad.com/media-mentions/study-mobile-ad-geo-targeting-and-geo-conquesting-on-the-rise/

21. N. Fong, Z. Fang, and X. Luo, "Geo-conquesting: Competitive locational targeting of mobile promotions," *Journal of Marketing Research* 52 (5), 2015: 726–735.

22. J.-P. Dubé, Z. Fang, N. Fong, and X. Luo, "Competitive price targeting with smartphone coupons," working paper, University of Chicago, 2016.

23. http://www.msi.org/reports/competitive-price-targeting-with-smartphone-coupons/

24. S. Hui, J. Inman, Y. Huang, and J. Suher, "The effect of in-store travel distance on unplanned spending: Applications to mobile promotion strategies," *Journal of Marketing* 77 (2), 2013: 1–16.

25. S. Hui, J. Inman, Y. Huang, and J. Suher, "Estimating the effect of travel distance on unplanned spending: Applications to mobile promotion strategies," *Journal of Marketing* 77 (March), 2013: 1–16.

26. S. Hui, J. Inman, Y. Huang, and J. Suher, "Deconstructing the 'first moment of truth': Understanding unplanned consideration and purchase conversion using in-store video tracking," *Journal of Marketing Research* 50 (4), 2013: 445–462.

27. S. Nirjon, J. Lui, G. DeJean, B., Priyantha, J. Yuzhe, and T. Hart, "COIN-GPS: Indoor Localization from Direct GPS Receiving." http://research.microsoft.com/en-US/people/liuj/coingps-release.pdf.

28. https://blog.passkit.com/how-to-use-eddystone-url-and-what-is-meant-by-no-need-for-an-app/

29. Proximity Marketing in Retail Q1 2016 Report.

30. D. Molitor, P. Reichart, M. Spann, and A. Ghose, "Measuring the effectiveness of beacon-based mobile store-level promotions," working paper, New York University, 2016.

31. Proximity Marketing in Retail Q1 2016 Report.

32. http://www.geomarketing.com/rite-aid-preps-one-of-the-largest-beacon-activations-across-all-4600-stores

33. http://blog.beaconstac.com/2016/03/5-location-based-marketing-technologies-agencies-should-leverage-in-2016/

34. http://www.geomarketing.com/macys-to-test-beacon-messages-outside-app-explore-retargeting

35. http://streetfightmag.com/2015/12/22/10-top-location-based-marketing-campaigns-of-2015/

36. S. Daurer, D. Molitor, M. Spann, and P. Manchanda, "How, where, and when consumers search on mobile," working paper, Marketing Science Institute, 2016.

37. A. Ghose and S. P. Han, "An empirical analysis of user content generation and usage behavior on the mobile Internet," *Management Science* 57 (9), 2011: 1671–1691.

38. https://www.placeable.com/blog/post/geo-targeting-geo-conquesting/

Chapter 6

1. http://ftw.usatoday.com/2015/12/howard-cosell-john-lennon-monday-night-football-video

2. http://www.newyorker.com/news/sporting-scene/baseball-and-bin-laden

3. SOASTA Survey: What App Do You Check First in the Morning? https://www.soasta.com/press-releases/soasta-survey-what-app-do-you-check-first-in-the-morning/

4. http://media-cmi.com/downloads/Sixty_Years_Daily_Newspaper_Circulation_Trends_050611.pdf

5. http://www.statista.com/statistics/183422/paid-circulation-of-us-daily-newspapers-since-1975/

6. http://www.businessinsider.com/ad-viewability-is-a-major-problem-on-mobile-devices-2016-4?IR=T

7. http://uk.businessinsider.com/the-ad-viewability-problem-and-solutions-2014-12

8. I. Chap, "How to master mobile advertising." http://www.freshbusinessthinking.com/business_advice.php?CID=18&AID=11090&PGI D=1.

9. R. Deiss, "The best time of day for mobile ad campaigns." http://drivingtraffic.com/the-best-time-of-day-for-mobile-ad-campaigns.

10. "Cracking the emerging markets report," Upstream Systems. http://www.upstreamsystems.com

11. B. Baker, Z. Fang, and X. Luo, "Hour-by-hour sales impact of mobile advertising," working paper, Social Science Research Network, 2014.

12. eMarketer 2012. Mobile banners continue to boast high click rates. http://www.emarketer.com/Article/Mobile-Banners-Continue-Boast-High-Click-Rates/1009299

13. https://nclac.wordpress.com/2012/09/18/4228/

14. J. Engel and R. Blackwell, *Consumer Behavior* (Dryden, 1982).

15. V. S. Ramaswamy and S. Namakumari, *Marketing Management*, fourth edition (Macmillan India).

16. http://www.mobilemarketer.com/cms/news/software-technology/23049.html

17. J. Zhang and L. Krishnamurthi, "Customizing promotions in online stores," *Marketing Science* 23 (4), 2004: 561–578.

18. K. Stilley, J. Inman, and K. Wakefield, "Spending on the fly: Mental budgets, promotions, and spending behavior," *Journal of Marketing* 74 (3), 2010: 34–47.

19. P. Danaher, M. Smith, K. Ranasinghe, and T. Danaher, "Where, when, and how long: Factors that influence the redemption of mobile phone coupons," *Journal of Marketing Research* 52 (5), 2015: 710–725.

20. J. Inman and L. McAlister, "Do coupon expiration dates affect consumer behavior?" *Journal of Marketing Research* 31 (3), 1994: 423–428.

21. A. Mishra and H. Mishra, "We are what we consume: The influence of food consumption on impulsive choice," *Journal of Marketing Research* 47(11), 2010: 1129–1137.

22. P. Kotler, *Marketing Management* (Prentice-Hall, 2002).

23. Z. Fang, B. Gu, X. Luo, and Y. Xu, "Contemporaneous and delayed sales impact of location based mobile promotions," *Information Systems Research* 26 (3), 2015: 552–564.

24. A. Ghose, A. Goldfarb, and S. Han, "How is the mobile Internet different? Search Costs and Local Activities," *Information Systems Research* 24 (3), 2013: 613–631.

25. N. Liberman and Y. Trope, "The psychology of transcending the here and now," *Science* 322 (5905), 2008: 1201–1205.

26. That experiment occurred over a three-day period during the last weekend of August 2012. They conducted the experiment in cooperation with an international chain of cinemas (IMAX theaters). For more information, see X. Luo, M. Andrews, Z. Fang, and C. Phang, "Mobile targeting," *Management Science* 60 (7), 2015: 1738–1756.

27. J. Zhang and L. Krishnamurthi, "Customizing promotions in online stores," *Marketing Science* 23 (4), 561–578.

28. Y. Trope and N. Liberman, "Construal-level theory of psychological distance," *Psychological Review* 117 (2), 2010: 440–463.

29. https://en.wikipedia.org/wiki/AIDA_(marketing)

30. A. Ghose, P. Singh, and V. Todri, "Trade-offs in online advertising: Modeling and measuring advertising effectiveness and annoyance dynamics," working paper, New York University, 2016.

31. For a good framework in which to think about these kinds of choices for firms, see M. Andrews, J. Goehring, S. Hui, J. Pancras, and L. Thornswood, "Mobile promotions: A framework and research directions," *Journal of Interactive Marketing* 34, May 2016: 15–24.

Chapter 7

1. A. Ghose and S. Yang, "An empirical analysis of search engine advertising: Sponsored search in electronic markets," *Management Science* 55 (10), 2009: 1605–1622.

2. A. Ghose, P. Ipeirotis, and B. Li, "Designing ranking systems for hotels on travel search engines by mining user-generated and crowd-sourced content," *Marketing Science* 31 (3), 2012: 493–520.

3. A. Ansari and C. Mela, "E-customization," *Journal of Marketing Research* 40 (2), 2003: 131–145.

4. http://screenwerk.com/2015/05/11/data-suggest-that-local-intent-queries-nearly-half-of-all-search-volume/

5. https://webmasters.googleblog.com/2016/11/mobile-first-indexing.html?m=1

6. https://searchenginewatch.com/sew/news/2395764/does-yelp-really-matter

7. A. Ghose, A. Goldfarb, and S. Han, "How is the mobile Internet different? Search costs and local activities," *Information Systems Research* 24 (3), 2013: 613–631.

8. J. Nunamaker, L. Applegate, and B. Konsynski, "Facilitating group creativity: Experience with a group decision support system," *Journal of Management Information Systems* 3 (4), 1987: 5–19.

9. Scientists refer to this need to think harder as "cognitive load."

10. M. Chae and J. Kim, "Do size and structure matter to mobile users? An empirical study of the effects of screen size, information structure, and task complexity on user activities with standard Web phones," *Behavior & Information Technology* 23 (3), 2004: 165–181.

11. S. Sweeney and F. Crestani, "Effective search results summary size and device screen size: Is there a relationship?" *Information Processing Management* 42 (4), 2006:1056–1074.

12. M. J. Albers and L. Kim, "User Web browsing characteristics using palm handhelds for information retrieval," in *Proceedings of IEEE Professional Communication Society International Professional Communication Conference and Proceedings of the 18th Annual ACM International Conference on Computer Documentation: Technology & Teamwork* (IEEE, 2000).

13. H. Davison and C. Wickens, "Rotocraft hazard cueing: The effects on attention and trust," technical report ARL-99–5/NASA-99–1, University of Illinois Aviation Research Lab, 1999.

14. The users in the study chose 4,557 coupons. Based on the number of impressions (354,662), this translates to a click-rate of 1.28 percent. Group 3 (ranking by distance, but without distance information) had the highest CTR at 1.65 percent, whole Group 4 (ranking randomly, without distance information) had the lowest CTR. Regardless of the group, these CTRs are considerably higher than those of mobile banner ads, with range from 0.1 percent to 0.4 percent (eMarketer 2014b).

15. A. Ghose, P. Singh, and V. Todri, "Trade-offs in online advertising: Modeling and measuring advertising effectiveness and annoyance dynamics," working paper, New York University, 2016.

16. https://blog.kissmetrics.com/app-store-optimization/

Chapter 8

1. http://www.wsj.com/articles/singapore-is-taking-the-smart-city-to-a-whole-new-level-1461550026

2. https://www.techinasia.com/singapore-advanced-surveillance-state-citizens-mind

3. Ibid.

4. R. Schmitt, "Density, health, and social disorganization," *Journal of the American Institute of Planners* 32 (1), 1966: 38–40.

5. J. Collette and S. D. Webb, "Urban density, household crowdedness and stress reactions," *Journal of Sociology* 12 (3), 1976: 184–191.

6. D. Sherrod, "Crowdedness, perceived control, and behavioral aftereffects," *Journal of Applied Social Psychology* 4 (2), 1974: 171–186.

7. W. Griffitt and R. Veitch, "Hot and crowded: Influence of population density and temperature on interpersonal affective behavior," *Journal of Personality and Social Psychology* 17 (1), 1971: 92–98.

8. P. Zimbardo, "The human choice: Individuation, reason, and order versus deindividuation, impulse, and chaos," in *Nebraska Symposium on Motivation 1969*, ed. W. Arnold and D. Levine (University of Nebraska Press).

9. M. Hui and J. Bateson, "Perceived control and the effects of crowding and consumer choice on the service experience," *Journal of Consumer Research*, 18, 1991: 174–184.

10. J. Brehm, *A Theory of Psychological Reactance* (Academic Press, 1966); R. Wicklund, *Freedom and Reactance* (Erlbaum, 1974).

11. A. Maeng, R. Tanner, and D. Soman, "Conservative when crowded: Social crowdedness and consumer choice," *Journal of Marketing Research* 50 (6), 2013: 739–752.

12. G. Harrell, M. Hutt, and J. Anderson, "Path analysis of buyer behavior under conditions of crowdedness," *Journal of Marketing Research* 17(1), 1980: 45–51.

13. R. Sommer, "Personal space," in *Encyclopedia of Human Relationships*, ed. H. Reis and S. Sprecher (SAGE, 2009).

14. B. McKenzie and M. Rapino, Commuting in the United States: 2009, American Community Survey Reports (US Census Bureau, 2011).

15. http://web.mta.info/nyct/facts/ridership/

16. http://www.railway-technology.com/features/featurethe-worlds-top-10-busiest-metros-4433827/

17. These numbers represent total ridership, not just weekday ridership. But data from New York City show that ridership falls off significantly on Saturdays and is only half as high on Sundays as on weekdays. Thus, we assume that the vast majority of these rides take place on weekdays.

18. M. Flegenheimer, "Wi-Fi and cellphone service on subway trains? M.T.A. leader says it may happen," *New York Times*, September 17, 2013. http://www.nytimes.com/2013/09/18/nyregion/mta-plans-wi-fi-and-phone-service-on-subway-trains.html

19. http://www.msi.org/articles/crowded-subways-boost-mobile-ad-response/

20. M. Andrews, X. Luo, Z. Fang, and A. Ghose, "Mobile ad effectiveness: Hypercontextual targeting with crowdedness," *Marketing Science* 35 (2), 2016: 218–233.

21. http://phys.org/news/2014-02-hong-kong-metro-seats-scrapped.html

22. Of the 10,690 mobile users who received a text message, 334 replied and purchased the promoted service.

23. D. Molitor, P. Reichart, M. Spann, and A. Ghose, "Measuring the effectiveness of location based advertising: A randomized field experiment," working paper, New York University, 2014.

24. The actual numbers per square meter were 1.96, 4.02, and 4.97. I have rounded them to full integers for simplicity's sake.

25. D. Stokols, "On the distinction between density and crowding: Some implications for future research," *Psychological Review* 79 (3), 1972: 275–277.

26. S. Milgram, "The experience of living in cities," *Science* 167 (3924), 1970: 1461–1468.

27. http://www.msi.org/articles/crowded-subways-boost-mobile-ad-response/

28. G. Evans and R. Wener, "Crowdedness and personal space invasion on the train: Please don't make me sit in the middle," *Journal of Environmental Psychology* 27 (1), 2007: 90–94.

29. http://indianexpress.com/article/technology/tech-news-technology/googles-free-wi-fi-at-railway-stations-is-being-used-by-2-mn-people-in-india-sundar-pichai-2942966/

30. R. Novaco, D. Stokols, and L. Milanesi, "Objective and subjective dimensions of travel impedance as determinants of commuting stress," *American Journal of Community Psychology* 18 (2), 1990: 231–257; M. Schaeffer, S. Street, J. Singer, and A. Baum, "Effects of control on the stress reactions of commuters," *Journal of Applied Social Psychology* 18 (11), 1988: 944–957; D. Kahneman, A. Krueger, D. Schkade, N. Schwarz, and A. Stone, "A survey method for characterizing daily life experience: The day reconstruction method," *Science* 306 (5702), 2004: 1776–1780.

31. D. Molitor, P. Reichart, M. Spann, and A. Ghose, "Measuring the effectiveness of location based advertising: A randomized field experiment," working paper, New York University, 2014. .

32. https://www.informs.org/About-INFORMS/News-Room/Press-Releases/Crowdedness-and-Mobile-Apps

33. eMarketer 2014.

Chapter 9

1. http://www.imediaconnection.com/articles/ported-articles/red-dot-articles/2015/jan/the-next-phase-of-location-based-advertising/

2. R. Sen, Y. Lee, K. Jaayarajah, A. Mishra, and R. Balan, "GruMon: Fast and accurate group monitoring for heterogeneous urban spaces," in *Proceedings of the 12th ACM Conference on Embedded Network Sensor Systems* (ACM, 2014).

3. Relevant references include the following: J. Howard, and J. Sheth, *The Theory of Buyer Behavior* (Wiley, 1969); D. Court, D. Elzinga, S. Mulder, and O. J. Vetvik, "The consumer decision journey," *McKinsey Quarterly*, June 2009.

4. J. Bettman, M. Luce, and J. Payne, "Constructive consumer choice processes," *Journal of Consumer Research* 25 (3), 1998: 187–217.

5. J. Engel and R. Blackwell, *Consumer Behavior* (Dryden, 1982).

6. V. Ramaswamy and S. Namakumari, *Marketing Management*, fourth edition (Macmillan India).

7. A. Ghose, B. Li, and S. Liu, "Mobile targeting using customer trajectory patterns," working paper, New York University, 2016.

8. Mobile showrooming is the practice of searching for the best prices or reading reviews on your smartphone even as you shop in the brick-and-mortar world.

9. See http://lifehacker.com/how-retail-stores-track-you-using-your-smartphone-and-827512308. The *Times* story, titled "Attention, shoppers: Store is tracking your cell," was published on July 14, 2013. .

10. P. Underhill, *Why We Buy: The Science of Shopping* (Simon & Schuster, 1999).

11. http://minneanalytics.org/shoppers-on-the-move/

12. http://www.latimes.com/business/la-fi-retail-heat-mapping-20160923-snap-story.html

13. https://www.dlapiper.com/en/abudhabi/insights/publications/2016/04/law-a-la-mode-issue-19/in-store-big-data-analytics/

14. http://www.dispatch.com/content/stories/business/2015/11/09/tech-firm-uses-data-from-video-to-help-store-owners-drive-sales.html

15. https://www.theguardian.com/technology/datablog/2014/jan/10/how-tracking-customers-in-store-will-soon-be-the-norm

16. L. Radaelli, D. Sabonis, H. Lu, and C. Jensen, "Identifying typical movements among indoor objects—Concepts and empirical study," presented at IEEE 14th International Conference on Mobile Data Management, 2013.

17. http://www.govtech.com/dc/articles/How-Can-Location-Based-Services-Overhaul-Local-Government.html

Chapter 10

1. T. Mills, "Some hypotheses on small groups from Simmel," *American Journal of Sociology* 63 (6), 1958: 642–650.

2. R. Zajonc, "Social facilitation," *Science* 149 (3681), 1965: 269–274.

3. S. Roper. and C. La Niece, "The importance of brands in the lunch-box choices of low-income British school children," *Journal of Consumer Behavior* 8 (2–3): 2009: 84–99.

4. X. Luo, "How does shopping with others influence impulsive purchasing?" *Journal of Consumer Psychology* 15 (4), 2005: 288–294.

5. K. Didem, J. Inman, and J. Argo, "The influence of friends on consumer spending: The role of agency—communion orientation and self-monitoring," *Journal of Marketing Research*. 48 (4), 2011: 741–754.

6. https://blog.crew.co/secret-science-of-shopping/

7. M. Andrews, X. Luo, Z. Fang, and A. Ghose, "Mobile ad effectiveness: Hyper-contextual targeting with crowdedness," *Marketing Science* 35 (2), 2016: 218–233.

8. D. Kollat and R. Willet, "Is impulse purchasing really a useful concept for marketing decisions?" *Journal of Marketing* 33 (1), 1969: 79–83.

9. R. Fisher and R. Simmons, "Smartphone interruptibility using density-weighted uncertainty sampling with reinforcement learning," in *Proceedings of the 2011 10th International Conference on Machine Learning and Applications and Workshops* (IEEE Computer Society, 2011).

10. Mills, "Some Hypotheses on Small Groups from Simmel."

11. A. Ghose, B. Li, and S. Liu, "Mobile advertising and real-time social dynamics: Evidence from a randomized field experiment," working paper, New York University, 2015.

12. A. Ghose, B. Li, and S. Liu, "Digitizing offline shopping behavior towards mobile marketing," in proceedings of 2015 International Conference on Information Systems, Fort Worth.

13. Ibid..

14. R. Hu and P. Pu, "Enhancing collaborative filtering systems with personality information," in *Proceedings of the Fifth ACM Conference on Recommender Systems* (ACM, 2011).

15. P. Rentfrow and S. Gosling, "The do re mi's of everyday life: The structure and personality correlates of music preferences," *Journal of Personality and Social Psychology* 84 (6), 2003: 1236–1256.

16. C.-F. Lin, "Segmenting customer brand preference: Demographic or psychographic," *Journal of Product & Brand Management* 11 (4), 2002: 249–268.

17. J. Chen, E. Haber, R. Kang, G. Hsieh, and J. Mahmud, "Making use of derived personality: The case of social media ad targeting," presented at Ninth International AAAI Conference on Web and Social Media.

18. H. Baumgartner, "Toward a personology of the consumer," *Journal of Consumer Research* 29 (2), 2002: 286–292.

19. P. Adamopoulos, A. Ghose, and V. Todri, "Estimating the impact of user personality traits on word-of-mouth: Textmining microblogging platforms," working paper, New York University, 2016.

20. For example, on Twitter highly accurate results could be obtained from only about 364 tweets.

21. http://www.wsj.com/articles/startups-see-your-face-unmask-your-emotions-1422472398

22. http://www.latimes.com/business/la-fi-retail-heat-mapping-20160923-snap-story.html

Chapter 11

1. http://www.nytimes.com/1999/10/28/business/variable-price-coke-machine-being-tested.html

2. http://www.ugamsolutions.com/blog/lessons-from-a-smart-vending-machine-when-is-it-ok-to-raise-prices

3. H. Simon, F. Bilstein, and F. Luby, *Manage for Profit, Not For Market Share: A Guide to Greater Profits in Highly Contested Markets* (Harvard Business School Press, 1996).

4. http://knowledge.wharton.upenn.edu/article/the-promise-and-perils-of-dynamic-pricing/

5. http://brettmcg.mlblogs.com/2014/10/02/quick-notes-on-the-weather/

6. http://adage.com/article/media/summer-tv-hits-misses/294330/

7. http://marketingland.com/seasonality-mobile-device-usage-warmer-weather-tempers-tech-95937

8. M. Persinger and B. Levesque, "Geophysical variables and behavior: XII. The weather matrix accommodates large portions of variance of measured daily mood," *Perceptual and Motor Skills* 57 (3), 1983: 868–870.

9. D. Hirshleifer and T. Shumway, "Good day sunshine: Stock returns and the weather," *Journal of Finance* 58 (3), 2003: 1009–1032.

10. C. Li, A. Reinaker, C. Zhang, and X. Luo, "Weather and mobile purchases: 10-million-user field study," working paper, Social Science Research Network, 2015.

11. J. Aaker and A. Lee, "Understanding regulatory fit," *Journal of Marketing Research* 43 (1), 2006: 15–19.

12. D. Molitor, P. Reichhart, and M. Spann, "Location-based advertising: Measuring the impact of context-specific factors on consumers' choice behavior," working paper, LMU Munich, 2014.

13. http://adexchanger.com/mobile/adtheorent-ceo-why-weather-affects-mobile-ctr-rates/

14. https://www.marketingweek.com/2014/02/05/how-the-weather-affects-marketing/

15. https://www.marketingweek.com/2014/02/05/how-the-weather-affects-marketing/

16. http://www.weatherunlocked.com/blog/2016/may/weather-targeting-for-adwords-the-ultimate-guide

17. http://www.mmaglobal.com/case-study-hub/case_studies/view/31812

18. https://www.marketingweek.com/2014/02/05/how-the-weather-affects-marketing/

19. http://www.fastcompany.com/3015821/fast-feed/how-the-weather-channel-predicts-what-youll-buy

20. http://www.psfk.com/2012/07/vending-machine-prices-change-by-temperature.html

21. http://www.nber.org/papers/w18212.pdf?new_window=1

22. http://www.weather.gov/buf/lake1415_stormb.html

23. http://oregonstate.edu/ua/ncs/archives/2002/dec/osu-program-nations-top-service-weather-channel-signs

24. K. Rosman, "Weather Channel now also forecasts what you'll buy," *Wall Street Journal*, August 14, 2013.

25. http://www.weatherunlocked.com/blog/2016/may/weather-targeting-for-adwords-the-ultimate-guide.

26. C. Li, A. Reinaker, Z. Cheng, and X. Luo, "Weather and mobile purchases: 10-million-user field study," working paper, Social Science Research Network, 2015.

27. http://www.weatherunlocked.com/blog/2016/may/weather-targeting-for-adwords-the-ultimate-guide

28. https://brain.do/blog/mobile-marketing-is-changing-with-the-weather/

Chapter 12

1. http://www.emarketer.com/Article/Tablet-Users-Surpass-1-Billion-Worldwide-2015/1011806

2. L. Einav, J. Levin, I. Popov, and N. Sundaresan, "Growth, adoption, and use of mobile E-commerce," *American Economic Review* 104 (5), 2014: 489–494.

3. http://marketingland.com/seasonality-mobile-device-usage-warmer-weather-tempers-tech-95937

4. D. Grewal, Y. Bart, M. Spann, and P. Zubcsek, "Mobile advertising: A framework and research agenda," *Journal of Interactive Marketing* 34, May 2016: 3–14.

5. Based on a conversation with Ramasubramaniya Raja and Anshuman Chaudhary.

6. P. Kannan, W. Reinartz, and P. Verhoef, "The path to purchase and attribution modeling: Introduction to special section." *International Journal of Research in Marketing* 33 (3), 2016: 449–456.

7. A. Ghose and V. Todri, "Toward a digital attribution model: Measuring the impact of display advertising on online consumer behavior," *MIS Quarterly* 40 (4), 2016: 889–910.

8. http://www.msi.org/reports/the-role-of-mobile-devices-in-the-online-customer-journey/

9. A. Ghose, A. Goldfarb, and S. Han, "How is the mobile Internet different? Search costs and local activities," *Information Systems Research* 24 (3), 2013: 613–631.

10. G. Sterling, "Saying a third of mobile searches are local, Google brings 'promoted pins' to maps." http://searchengineland.com/saying-third-mobile-searches-local-google-brings-new-ads-maps-250278

11. http://www.adweek.com/news/technology/foursquares-potentially-game-changing-new-tool-can-measure-foot-traffic-generated-digital-ads-169785

12. http://www.adweek.com/news/technology/foursquares-potentially-game-changing-new-tool-can-measure-foot-traffic-generated-digital-ads-169785/

13. https://techcrunch.com/2016/02/22/attribution-by-foursquare/

14. http://www.businessinsider.com/comscore-begins-tracking-offline-foot-traffic-and-sales-2016-6?r=US&IR=T&IR=T

15. http://www.mobilemarketer.com/cms/news/advertising/23058.html

16. https://developers.facebook.com/docs/analytics/training

17. Grewal et al., "Mobile advertising: A framework and research agenda."

18. http://amobee.com/amobee-impact-mobile-ad-formats-drive-consumer-engagement-rates/

19. For examples of native ads, see "Native advertising examples: 5 of the best (and worst)" at http://www.wordstream.com/blog/ws/2014/07/07/native-advertising-examples.

20. https://www.mobileads.com/blog/best-mobile-ad-formats-sizes-display-ad-campaigns/

21. http://www.wsj.com/articles/advertisers-try-new-tactics-to-break-through-to-consumers-1466328601?mod=e2tw

22. http://adage.com/article/digitalnext/effective-native-ads-a-solution-ad-blockers/302476/

23. http://knowledge.wharton.upenn.edu/article/indias-leading-fashion-e-tailer-abandoned-app-strategy/

24. http://blog.adspruce.com/how-mobile-web-beats-mobile-app/

25. https://www.clickz.com/m-commerce-has-the-mobile-web-finally-won/103284/

26. https://www.clickz.com/m-commerce-has-the-mobile-web-finally-won/103284/, 2016.

27. S. Aral and P. Dhillon, "Digital paywall design: Effects on subscription rates and cross-channel demand," working paper, Massachusetts Institute of Technology, 2016. .

28. http://www.inmobi.com/blog/2016/02/18/the-marketers-dilemma-mobile-web-vs-app

29. https://blog.appboy.com/deep-linking-can-double-conversion-increase-retention-boost-engagement/

30. T. Sun, L. Shi, S. Viswanathan, and E. Zheleva, "Motivating mobile app adoption: Evidence from a randomized field experiment," working paper, University of Maryland, 2016.

31. L. Odden, "B2B Mobile Marketing for Demand Generation? Yes! Examples and Quick Tips." http://www.toprankblog.com/2015/07/b2b-mobile-marketing/

32. http://www.emarketer.com/Report/Six-B2B-Mobile-Marketing-Trends-2016-With-More-Mobile-Workplace-Budgets-Tactics-Must-Follow/2001718

33. P. Hendrix, The Engagement Stack. http://knowledge.brandify.com/the-engagement-stack/

Chapter 13

1. http://www.nytimes.com/2016/08/03/technology/china-mobile-tech-innovation-silicon-valley.html

2. George S. Yip and Bruce McKern, *China's Next Strategic Advantage: From Imitation to Innovation* (MIT Press, 2016). .

Chapter 14

1. Y. Dang, A. Ghose, X. Guo, and B. Li, "Empowering patients with diabetes using smart mobile health platform: Evidence from a randomized trial," working paper, New York University, 2016.

2. http://www.digitaltrends.com/wearables/smart-clothing-is-the-future-of-wearables/

3. http://news.wgbh.org/post/watch-new-tool-analyzes-your-facial-expressions

4. https://techcrunch.com/2016/05/25/affectiva-raises-14-million-to-bring-apps-robots-emotional-intelligence/

5. http://www.wsj.com/articles/startups-see-your-face-unmask-your-emotions-1422472398

6. http://mobilemarketingmagazine.com/samsung-warns-smart-tvs-may-snoop-on-your-conversations/

7. http://www.wsj.com/articles/if-your-teacher-sounds-like-a-robot-you-might-be-on-to-something-1462546621

8. http://www.mindshareworld.com/ireland/news/facebook-chatbots-and-future-messaging-apps

9. http://wearesocial.com/uk/blog/2016/03/the-messaging-app-market-and-its-future-potential

10. http://www.cmo.com/features/articles/2015/2/26/five_innovative_mobi.html

11. http://wearesocial.com/uk/blog/2016/03/the-messaging-app-market-and-its-future-potential

12. http://www.nytimes.com/2016/08/03/technology/china-mobile-tech-innovation-silicon-valley.html.

13. http://thenextweb.com/asia/2014/05/30/brands-can-now-set-up-shop-inside-chinese-messaging-app-wechat/#gref

14. http://technode.com/2016/09/22/wechats-app-within-app-free-last-endless-installing-deleting/

15. http://www.wsj.com/articles/the-future-of-texting-e-commerce-1451951064

16. http://www.mindshareworld.com/ireland/news/facebook-chatbots-and-future-messaging-apps

17. http://www.mobilemarketer.com/cms/news/research/23127.html

18. http://www.thedrum.com/news/2016/01/11/smartest-way-forward-marketers-smart-home

19. https://www.theinformation.com/apple-opening-siri-developing-echo-rival

20. D. Hoffman and T. Novak, "Emergent experience and the connected consumer in the smart home assemblage and the Internet of Things," working paper, Social Science Research Network, 2015.

21. http://mobilemarketingmagazine.com/97761-2/

22. http://www.theverge.com/2016/1/5/10711914/ford-smart-home-connectivity

23. http://mashable.com/2016/07/06/volkswagen-lg-connected-car-agreement/#5zinx7Naimq4

24. http://mobilemarketingmagazine.com/97761-2/

25. http://www.smartinsights.com/mobile-marketing/5-mobile-marketing-trends-that-will-rule-in-2016/

26. http://www.inc.com/walter-chen/pok-mon-go-is-driving-insane-amounts-of-sales-at-small-local-businesses-here-s-h.html

27. http://adage.com/article/digital/sponsored-locations-coming-pok-mon-a-cost-visit-basis/304952/

Epilogue

1. Pebble has developed a service called "the Concierge" whereby users can get anything they want simply by talking to the service. This voice interface is instrumental in enhancing the overall experience of its users with wearable technology.

2. The Federal Bureau of Investigation wanted Apple to create new software that would enable the FBI to unlock an iPhone 5C it had recovered from one of the shooters in the December 2015 terrorist attack in San Bernardino, California.

3. http://www.wsj.com/articles/apples-new-tech-will-peek-at-user-habits-without-violating-privacy-1466069400

4. http://www.techinsider.io/apple-is-beefing-up-artificial-intelligence-2016-5

5. https://www.bostonglobe.com/arts/2016/06/05/why-our-location-histories-are-glimpse-into-future/OqFyyVTy1sWWBJRDguvX5K/story.html

6. http://newsroom.fb.com/news/2014/06/making-ads-better-and-giving-people-more-control-over-the-ads-they-see/

7. http://www.aboutads.info/choices/

8. https://www.bostonglobe.com/arts/2016/06/05/why-our-location-histories-are-glimpse-into-future/OqFyyVTy1sWWBJRDguvX5K/story.html

9. http://crackberry.com/emergency-why-cellular-data-better-voice

10. http://pilipinas911.com/main/

11. http://www.statista.com/statistics/274774/forecast-of-mobile-phone-users-worldwide/

12. http://www.rmai.in/ejournal/national-international-trend/5-how-smartphones-are-penetrating-deeper-in-rural-india

13. S. Cole and A. Fernando, "'Mobile'izing agricultural advice: Technology adoption, diffusion and sustainability," working paper, Harvard Business School, 2016.

14. http://www.strategy-business.com/article/14196?gko=72c80

15. R. Reich, *The Future of Success: Working and Living in the New Economy* (Vintage, 2002).

16. http://www.bbc.com/news/science-environment-31450389

17. http://www.forbes.com/sites/jaysondemers/2014/02/10/2014-is-the-year-of-digital-marketing-analytics-what-it-means-for-your-company/#1b0494806619

18. http://www.wsj.com/articles/can-you-sue-the-boss-for-making-you-answer-late-night-email-1432144188

19. http://www.emarketer.com/Report/Six-B2B-Mobile-Marketing-Trends-2016-With-More-Mobile-Workplace-Budgets-Tactics-Must-Follow/2001718

Index

Accessibility, 26
Ad blockers, 36–38, 81, 165, 172
Ad exposures, 68, 81, 82, 90–92, 103, 108, 164
Ad Preferences tool, 6, 7, 37
Advertising
  annoyance and, 4–6, 9, 33–37, 91, 96, 109, 125, 128
  format of, 171, 172, 178
  framing of, 83, 84, 93
  granular, 80–82
  location-based, 28, 54, 60–64, 68, 69, 101–103, 106, 119, 126, 130
  mobile, 27, 28
  non-viewability of, 164, 165
  pull, 63, 67, 78, 101–104
  push, 62, 78, 135
  spending on, 20, 21
  trajectory-based, 126–131, 134
  unsolicited, 35, 36
Agreeableness, 143
AIDA framework, 91
Always-on lifestyle, 1, 26
App indexing, 105
Apple, 197, 198
Apps vs. mobile websites, 173–177
Artificial intelligence technologies, 182, 186, 187
Avoidance behavior, 186

Banner ads, 171
Bar-code-scanning apps, 76, 77
Beacons, 73–78
Behavioral constraint theory, 112, 113
Behavioral targeting, 71, 72, 78
Blue-light specials, 9
Bluetooth Low Energy, 73–76
Brazil, 11, 18, 42, 75

Captive shoppers, 72, 73
Cars
  connected, 191, 192
  purchasing of, 153, 154
  self-driving, 191
  smart, 191, 192
Certainty, 34, 35
Charitable donations, 75, 76
Chatbots, 189
China, 11, 17-18, 42, 83, 176, 182, 188
Choice, 12, 38, 39
Choice overload, 38, 39
Collaborative economy, 27
Communication
  changes in, 25, 26
  two-way, 50
Commuters, 114–118, 191
Competitive locational targeting, 70, 71
Complementarities between devices, 161–163, 174
Concierge, smartphone as, 6
"Concierge feeling," 38

Conscientiousness, 143
Construal-level theory, 90
Consumer behavior
  changes in, 11, 12, 19, 25, 27
  contradictions in, 33–40
  groups and, 140
  modes and, 54, 55
  physical traces of, 30, 31
  similarities in, 2, 3
  social context and, 136, 137
  stages of purchase and, 87, 90–93, 123
  trajectory-based advertising and, 129
Consumers' personas, 48–51
Coupon redemption windows, 85–90, 93, 116–118
Crowdedness, 62, 107–118, 137, 186
Cruise ships, 47, 48

Data
  collected, 29
  as currency, 39, 40
  predictive models and, 44
  protection of, 24
  role of, 6, 7
  trust and, 3
  use of, 12
  value of offline, 127
  volume of, 29
Data breaches, 8, 40, 196, 197
Dayparting, 81
Decision paralysis, 39
Deep linking, 175, 176
Delayed sales effects, 87, 93
Demographics, 128, 129
Differential privacy, 197
Digital attribution, 163, 164, 177
Direction of travel, 67, 68
Discounts
  distance and, 65–68
  ranking and, 101–105
  social, 140, 141, 146
Diversification bias, 34

Downtime productivity, 26, 27
Dynamic pricing, 147, 148, 153

Eddystone technology, 74
Emerging technologies, 181, 182
Emotional range, 143
Event-based marketing, 52
Explorers (shoppers), 85, 127, 134
Extraversion, 143
Eye-tracking technology, 105, 106

Facebook, 6, 7, 37, 38, 188, 189, 198
Facial expressions, reading, 144, 186
Fear of missing out, 35–38
Focused shoppers, 85, 127, 134
Fraud, 21
Freedom, 38, 39

"Geeks," 201, 202
Geo-awareness, 61–65
Geo-conquesting, 64, 68–72, 77
Geo-fencing, 29, 64, 65, 68, 78
Geo-targeting, 64, 68, 71, 72, 77, 78, 89, 93
Germany, 18, 23, 37, 67, 74, 101, 151
Give-and-take relationships, 42, 57
Google, 197, 198
Guidance functions, 115

"Haircast," 152, 153
Heat mapping, 131
Hedonic products, 83, 84, 93
High-income customers, 55, 128, 141
Human–computer interaction, 137
Human trace data, 30

Immediate purchase behavior, 86, 87
Implants, 187, 196
India, 18, 23, 56, 114, 173
Indoor positioning technologies, 124, 125
Indoor targeting precision, 73–76
Information chunking, 100

Information overload, 39
Instant messaging, 187–189
Interdependencies, 165–168
Internet of Things, 182, 189–192
Interstitial ads, 171
In-the-moment marketing, 55, 56
IPhone, 21, 22

Leisure versus work time, 26
Local searches, 98
Location, 59–61
  data on, 24
  distance and discount and, 65–68
  engagement and, 72, 73
  geo-awareness and, 61–65
  geo-conquesting and, 69–72
  indoor targeting precision and, 73–76
  mobile showrooming and, 76, 77
  ranking and, 101–105
  redemption windows and, 87–90
  takeaways regarding, 78
Location-based services, 28, 29
Lyft, 28–29

Meeker, Mary, 21
Messaging technologies, 182, 187–189
Micro-climates, 154–158
Micro-moments, 55–58, 82–84, 93
Mobile devices
  adoption of, 17–20
  cross-channel use and, 165–168
  evolution of, 21, 22
  substitution and complementarities between, 161–163
Mobile ecosystem, 9
"Mobile first" attitude, 18
Mobile GDP, 18
Mobile immersion, 112, 113
Mobile indexing, 98
Mobile Location Analytics code of conduct, 132, 133
Mobile pull, 63, 78, 101–104
Mobile push, 62, 78, 135
Mobile showrooming, counteracting, 76–78
Mobile wallet services, 23
Mobile websites, apps vs., 173–177
"Modes," 48, 49, 54, 55
Mood, weather and, 149–151
Motivation, 51, 52
Multi-screen behavior, 160–163

Native ads, 171–173, 178
Nearbuy, 56
Newspaper circulation, 80, 81
Non-commuters, 114–118
Notice, data and, 12

Offline trajectory, 30, 31
Omni-Channel, 75, 158–160, 171, 177, 203
Openness, 143

Paradox of choice, 38, 39
Personality traits, 142–146
Personalization, 40
Phone calls, frequency of, 25, 26
Pokémon Go, 193
Position effect, 97
Predictive models, 44
Prevention framing, 150
Primacy effect, 96, 99, 100
Privacy, 39, 40, 132, 187, 190, 191, 196, 197
Programmatic marketing, 203
Proximity marketing, beacon-enabled, 73–76
Proximity solution providers, 23
Purchase funnel, 87, 90–93, 123, 165

Quick response codes, 68, 69

Radio-frequency identification, 131
Ranking, 97–106
Recency effect, 96

Reich, Robert, 201
Rich media messaging, 63

Saliency, 95–106, 185
Saliency tax, 99, 100
Schmidt, Eric, 18
Screen size, 99, 100, 105
Search engine optimization, 98, 99
Searches, mobile vs. PC, 168, 169
Semantics, 122
Sharing economy, 27
"Shrinks," 201, 202
Simon Personal Communicator, 21, 22, 181
Singapore, 107, 108
Smart homes, 189–192
Smart wallets, 192, 193
Social context, 136–140
Social dynamics, 54, 55, 135–146
Social impact of technology, 199–201
South Korea, 17, 18, 77, 114, 131, 165, 182, 188
Sponsored locations, 193
Spontaneity, 34, 35
State of mind, 54, 55
Substitution between devices, 161–163, 174
Subways, 109–114
Surge pricing, 153

Tablets, 161–163
Tech mix, 159, 160
Telecommuting, 26, 27
Temporal targeting, 89, 93
Texting, 25
Text-messaging campaigns, 62, 63
Time
  of day, 82–84
  granularity and, 80–82
  of offer validity, 85–87
  trajectory and, 121, 130
  of week, 84, 85
Trajectory, 30, 31, 53, 54, 120–127, 133
Trust, 3, 8

Uber, 28, 29
"Usuals," 52–54, 58
Utilitarian products, 83, 84, 93

Validity periods, 85–87
Variable pricing, 147
Velocity, 121, 122
Video ads, 171
Viewability, 81
Virtual reality/augmented reality, 192–194
Voice-recognition software, 187

Wanamaker's riddle, 159, 168, 177
Wearable technologies, 182–185, 192
Weather, 147–158
WeChat, 188
Welcome promotions, 75
Wi-Fi technology, 22, 23, 124
"Work from home" culture, 26
Work versus leisure time, 26